How to Time Travel

A Collection of Timely Short Stories

(Fun Facts & Theories on How to Travel Through Time)

Brandy Kerr

Published By **Zoe Lawson**

Brandy Kerr

All Rights Reserved

How to Time Travel: A Collection of Timely Short Stories (Fun Facts & Theories on How to Travel Through Time)

ISBN 978-1-77485-698-7

No part of this guidebook shall be reproduced in any form without permission in writing from the publisher except in the case of brief quotations embodied in critical articles or reviews.

Legal & Disclaimer

The information contained in this ebook is not designed to replace or take the place of any form of medicine or professional medical advice. The information in this ebook has been provided for educational & entertainment purposes only.

The information contained in this book has been compiled from sources deemed reliable, and it is accurate to the best of the Author's knowledge; however, the Author cannot guarantee its accuracy and validity and cannot be held liable for any errors or omissions. Changes are periodically made to this book. You must consult your doctor or get professional medical advice before using any of the suggested remedies, techniques, or information in this book.

Upon using the information contained in this book, you agree to hold harmless the Author from and against any damages, costs, and expenses, including any legal fees potentially resulting from the application of any of the information provided by this guide. This disclaimer applies to any damages or injury caused by the use and application, whether directly or indirectly, of any advice or information presented, whether for breach of contract, tort, negligence, personal injury, criminal intent, or under any other cause of action.

You agree to accept all risks of using the information presented inside this book. You need to consult a professional medical practitioner in order to ensure you are both able and healthy enough to participate in this program.

TABLE OF CONTENTS

Introduction 1

Chapter 1: The What Is Time Travel? 7

Chapter 2: Time Travel To The Past 23

Chapter 3: Technology: Current Technology 33

Chapter 4: Time Travel To The Future: ... 45

Chapter 5: Philosophical Understandings Of Time Travel: 53

Chapter 6: Paradoxes Of Time Travel 77

Chapter 7: Faster Than Light 86

Chapter 8: Time Dilation 130

Chapter 9: Major Scientific Study Launched 162

Conclusion 184

Introduction

One of the best benefits of the concept of parallel universes is that it could make time travel feasible. Why? Because it eliminates one of the more difficult complications associated with time travel, the so-called "Grandfather Paradox". This implies the possibility that traveling through time (into into the past) would not be possible since any person who travels backwards in time can perform actions that could alter the course of events in the near future. One of the most famous examples is someone going back to kill his or his grandfather. It's not possible, since If someone murdered their own grandfather, then the person wouldn't have ever existed. become a time traveler!

If our universe actually consists of billions and millions of millions of other universes it is possible for a person to travel backwards through time and land not on his own timeline or universe, but rather in an alternate universe, and one that is so close to the real world that it is nearly identical.

All the benefits of time travel actually could be realized. Even if our savage time traveler kills his grandfather from the past, it will be a no-brainer since the grandfather would be dead in a different 'side world', and lived his entire existence in the time traveller's world.

That's just one way in which time travel and the theory of parallel universes cross paths. One more paranormal scenario could be the ongoing ghosts. Consider this:

If someone dies, they immediately become an element in the history of that person's death. But, if their ghost is visible as a present-day entity, then the ghost is now part of both the past as well as the present. A fascinating way of dealing the issue is to put ghosts within a parallel universe. When a ghost is observed it is experiencing a glimpse of an alternate universe within the current moment.

It gets more complicated as it goes on. In this book, we don't care as much about the difficult maths and mind-boggling science, as well as the physical physics behind Parallel Universes and Time Travel as well as the reality that we find themselves in the middle of these phenomena nearly every every day!

Both are not only linked however, they also provide a myriad of other issues related to paranormal phenomena such as:

UFOs and UFO-related activities

Ghost and ghost-like phenomenon poltergeist

Cryptozoology (Bigfoot Shape-shifting Skinwalkers The Loch Ness monster)

Ancient monuments which act as gateways to other realms

Strange sightings of cities and landscapes

... and more. All of these are phenomena that will not disappear. They continue to occur, and science continues to deny all of it and the whole thing begs an logical and actually, a scientific explanation.

The truth is that the combination of time travel theory and an idea that the parallel universes could be real, it provides an amazing explanation for the above. It is no longer the case that millions of people need to be dismissed as unstable, delusional or insane. A practical model of time travel and the structure of parallel universes not just explain why a lot of normal and normal people are experiencing bizarre experiences, but puts a variety of 'paranormal ' phenomena into the realm of normal phenomena'.

In this book, we'll examine real-world cases published in the main media that strongly suggests that a variety of people from every walk of life have had experiences with time travel as well as parallel universes. often both simultaneously.

One of the more well-known UFO instances ever The British Roswell and an infamous observation of floating cities across China repeatedly (no joke intended) individuals are confronted by phenomena that are not currently explained beyond the parallel universe and time travel phenomena. In addition, this is the case with phenomena like the crop circles, the sightings of Bigfoot and bizarre encounters with shape shifters within the deserts in southwest American southwest, and bizarre experiences related to the stone structures, which include natural as well as man-made.

In the past, was considering time travel or access to parallel universes as something real needed to be so open-minded and could actually be thought of as being open-minded until the point that they were uninformed. We believe the accounts that you read here demonstrate that the pendulum is shifting to the side of what many people have reported on good-faith. They say that what they've experienced is true. In the meantime the typical dismissals made by those who are hard-bitten and material-science skeptical are becoming less believable each day.

It is a cliche that says:

If you call a man a pet once and then you ridicule him. Make him call an animal a thousand times, and then he'll begin barking.

One or two experiences with time travel, or a brief glimpse into an alternate universe could easily be dismissed as an unnatural event or an untrue abnormality of people who witnessed it. However, if these events happen repeatedly time again, and also to regular people with a healthy mind and good standing or to thousands of people simultaneously even the skeptical might have to begin barking.

We're sure to allow the possibility of being as skeptical as you want about these stories, but maintaining an open mind. NASA scientist Tom Campbell always recommends 'open-minded doubt' as the most effective method of examining subject matter in the fringe. However, Campbell adds that this is where the most significant breakthroughs in science have always been derived from the fringe. Campbell says we need to be able to be able to cope with uncertainty.

Good advice to take in when you are faced with the incredible stories you're about learn about within this novel.

Chapter 1: The What Is Time Travel?

The words used by scholar, scholar and politician Benjamin Franklin, "Time is money." In the equally acclaimed words of the essayist K. Bromberg, "Time is important. Make the most of it." The notion the notion that time can be a finite product has grown exponentially over the ages and we are often encouraged to use time in the best way that can be expected given that eventually it will run out and it is not as clear. The latest research in theoretical science is examining this narrow perspective on time, and scientists in the field are beginning to speculate about the possibility in time travel.

Since Einstein presented his theory regarding general relativity back in 1915 in which he claimed that time moves differently with frame references that are moving at different speed, the notion of time has changed dramatically. Although time was once considered to be an all-inclusive continuous, linear motion of existence however, today, physicists argue about the fundamental properties of time, such as whether it is constant or quantized and if it is unidirectional (just pushing forward) or directional (equipped to advance and then moving backwards). These baffling characteristics of time have led to the idea of time travel and a lot of physicists believe that it's theoretically possible.

To make time travel possible, the body needs to begin at the first point and move onto the nextone, or to a later date or in the past.

Time travel to the future that is to come is remarkably more comprehensible and more useful as compared to time travel in the past. In the past Einstein's relativity states that time changes in different ways for people who are moving at different velocity. When applied, this notion implies that if a person could accelerate to a speed that is similar to that of light, then the time of the body to move would change at a much slower pace with comparison to a still body.

This idea is frequently portrayed in the realm of hypothetical science by the twin conundrum. If Twin A was to stay on Earth and Twin B was in an rocket and traveled close to the speed of light, and then separated in space, and then pivoting and returning to Earth The reality of the situation will be apparent when the time is over is a factor for Twin A with respect to Twin B's point of reference as well Twin A will have matured more than Twin B. Although Twin B did nothing other than move at a high speed and is basically looking off for the future via the process of time dilation.

Time travel to the past uses the same principles of Einstein's relativitybut in a much more confusing manner. There are several ways where a human body could effectively reverse time. The first one is based on the rule mentioned above in Einstein's theory of relativity. Since frame references at higher speeds are slower in time, if an object was to move faster than light speed and reach the point at which light itself arrives and thus reach an event before we could observe it in the past and thereby bringing it to the past.

However, a different method for time travel in the past relies on another rule of thumb of Einstein's theory that the distinct gravitational fields draw frames in a surprising way, more so because more grounded gravitational fields make time move more slowly. As an example, similar to how the Earth has a gravitational pull on all the bodies that surround it, other bodies also possess a gravitational field which draws on the surrounding bodies. These pulls on gravity affect the course of time, but in different ways for all bodies within the gravitational field. In a similar way to the theory regarding time travel some scientists believe that time travel in the past can be achieved by using the wormhole.

A wormhole is a marvel that allows for a smooth transition between two focal points in space-time. Wormholes are similar to bridges: instead of traversing the length of the focuses that are between the two, a wormhole connects the two focuses. By connecting these two pieces of information, physicists suggest making one wormhole nearly the size of a massive gravitational field, and another that is near to a considerably smaller gravitational field to create a possible time machine.

Because time is synchronized within the wormhole, but not synchronized outside of the wormhole, the observer could enter the wormhole from the gravitationally thicker end at the time A and then exit the wormhole on the other end at B. But considering how the dense gravitational field is able to reverse time with respect to the field that is not as thick the observer is leaving at the moment A from the thicker end and has been in the thin field. The enlargement of this time is sufficient to trigger the development to reverse time relative to the area of gravitationally weakening.

Despite the fact that headway advancements have been created to support time travel, it's an untested possibility. The application of these theories in real tests requires a more skilled and extensive understanding of the science. Yet, to things that were considered to be pure fiction in the past 100 years the concept of time travel is an example of being one way in which science expands its boundaries and makes impossible things possible.

The concept of time travel refers to the notion of progress, (for instance, an individual) at specific times in time, which is practically comparable to the process of development between different points in space. It is typically done using an idea of a device known as the time machine, either as the vehicle of an entryway that connects impossible dates in time. Time travel is an enticing notion in the realm of fiction, but traveling to a arbitrary point in time has a restricted basis in physical science, usually only in conjunction with Einstein-Rosen spans or quantum mechanics. In a less specific sense, one-way travel to the future through using time dilation is an acknowledged incredible feat of relativistic science however, traveling to any significant "separation" requires rates that are similar to light, which isn't possible for human beings with today's technology. The concept was discussed in various prior works of fiction, but was not promoted until H. G. Wells in his novel of 1895 The Time Machine, which introduced the concept of time travel to the mainstream of society and is popular in science fiction.

A few myths from the past mention the possibility of advancing time. In Hindu mythology The Mahabharata tells the story that tells of the King, Raivata Kakudmi who travels to heaven to meet his creator Brahma and is shocked to discover when he returns home to Earth that a number of years have gone by. [2][3The Buddhist Pali Canon says the importance of time. The Payasi Sutta tells the story of one of his principal followers, Kumara Kassapa, who informs the skeptic Payasi that "In the heaven of Thirty Three Devas, time goes in a different direction, and people live for longer. "In the age of our century, 100 years is one day; twenty-four hours would have been lost to the devas. "[4]

A Japanese tale of "Urashima Taro",[5first told in the film Nihongi (720)[6tells the story of a young angler known as Urashima Taro who visits an undersea royal residence. After three daysin the palace, Urashima Taro goes home to his home town and is found 300 years later, after which he's been ignored and his house is a ruin and his family members have abandoned the family home.

When you sleep:

The Talmud describes Honi ha-M'agel who lies in bed for a long period of time, and wakes up to find his grandchildren are now grandparents, and that his family and friends have died. [7]

The novel of Louis-Sebastien Mercier from 1770, a romantic and L'An 2440 Reve s'il en the past ("The Year 2440 A Dream If There Was One") describes a character who finds himself in an alternate time of 2440. The novel is a cult classic, having been through a quarter-century since its publication in 1771, the novel focuses on the actions of an anonymous man who discusses with an academic companion about the shady crimes of Paris before he snoozes and is re-admitted to the future city of Paris.

The novel by Washington Irving "Tear Van Winkle" (1819) depicts the story of a man who has taken an unintentional nap for twenty years on the mountain and wakes up in a world in which he is not noticed while his partner has left and his daughter has developed. In addition, sleep is used as a means of traveling through time in H.G.

Wells' The Sleeper Awakes In the novel, the protagonist awakes after two hundred years of rest.

Time travel back to the past:

Similar to forward time travel backwards time travel also has an undetermined reason. Samuel Madden's Memoirs of the Twentieth Century (1733) is a sequence of letters sent by British ministers in 1998 and 1997 to negotiators before, passing on the religious and political future states. The storyteller receives the letters from his holy messenger Paul Alkon tells in his book Origins of Futuristic Fiction that "the first time-traveller in English writing is an angel of the gatekeeper."[8]:85 Madden does not clarify how the blessed messenger obtains the archives, however Alkon states that Madden "merits recognition as being the first person to play with the idea of time-travel as a quaint rare phenomenon that is reversed from the future, only to be discovered within the current."[8]:95-96

Mr. as well as Mrs. Fezziwig's movements in a dream, appeared to Scrooge through The Ghost from Christmas Past.

In 1836, Alexander Veltman distributed Predki Kalimerosa Aleksandr Filippovich Madedonskii (The Forefathers of Kalimeros: Alexander, child of Philip of Macedon) The novel has been referred to as the first original Russian sci-fi novel , and also the first novel to use time travel. The storyteller journeys back to ancient Greece by hippogriffs, is introduced to Aristotle and embarks on a journey together with Alexander the Great before returning to the 19th century.

In the sci-fi novel Far Boundaries (1951), the supervisor August Derleth claims that an early short story about time travel was "Feeling losing one's Coach An Anachronism" that was written by the Dublin Literary Magazine by a mysterious creator in 1838. When the storyteller is sitting close to a tree in search of an adviser to lead the storyteller out of Newcastle and into the future, he is taken to another time, spanning 1,000 years. He encounters an encounter with the Venerable Bede in an orthodox community and is able to reveal to him the advancements that are to come in the next several hundred years. However the tale never makes it clear if these experiences are real or a false dream. [11]:11-38

Charles Dickens' A Christmas Carol (1843) is one of the early depiction of time travel that spans both directions,[12as the main character, Ebenezer Scrooge, is transported to Christmases of the past as well as the future. In reality the story could be read as dreams, not time travel, since Scrooge experiences the different eras by being an observer, instead of the participant.

An even more clear example of reverse time travel is described in the well-known 1861 publication Paris avant les hommes (Paris before Men) by the French botanist and geologist Pierre Boitard, distributed after his death. In this tale the protagonist is taken to the past via the attraction of the "faltering evil force" (a French joke on Boitard's name) and is able to experience the presence of a Plesiosaur along with an apelike ancestor and befriend ancient creatures. [13]

Edward Everett Hale's "Hands Off" (1881) tells the story of an unknown being, possibly an individual who since passed away and who tamper with the antiquated Egyptian time by allowing Joseph to remain in subjugation. It could have been the primary story that was used to create the creation of a new history by time travel. [14]

One of the primary stories that demonstrates time travel through the use of a device is "The Clock that went backwards" written by Edward Page Mitchell. Mitchell's story appeared within the New York Sun in 1881. An unusual clock when wound, operates in reverse, and then transports people who are near to it to the past. The creator of the clock does not provide any information about the origins or characteristics for the particular clock. [16]

Enrique Gaspary Rimbau's El Anacronopete (1887)[17] may be the source of the primary story that features the construction of a vessel that could travel through time. [18The story of Andrew Sawyer has remarked that the tale "seems to represent the first research-based depiction of a time machine described this way" and includes "Edward Page Mitchell's novel The Clock That Moved Backward' (1881) is often portrayed as the first story about a time machine however I'm not sure that a clock can be considered a time machine in the same way. "[19[19.] H. G. Wells' The Time Machine (1895) introduced the concept of time travel using mechanical methods. [20]

Theory:

Certain theories, the most specifically, general relativity and unique suggests that reasonable geometry of spacetime or specific kinds of motion in space could allow time travel to the future and the past, If these kinds of geometries and movements were feasible. In papers that are specialized the majority of physicists remain a long way from the normal terminology that refers to "moving" and "going" in time. "Development" often refers to the change in the spatial location as the direction of the time shifts. Instead, they discuss the probability of shut timelike bends which are the world lines which form shut circles in spacetime, allowing frame references to go back to their previous. There aren't any answers to the questions of general relativity which describe spacetimes with closed timelike bends, such as for instance, Godel spacetime, however the physical validity of these structures is questionable.

Relativity suggests that, if someone were to travel far away from Earth in a relativistic manner and return, more time would have been lost in the direction of Earth in comparison to the explorer therefore, in this way it is recognized that relativity allows "going further into the future.. "Any theory that allows time travel could raise problems of causality. One example of a problem that includes causality concerns one of them being the "granddad trap 22" Think about the possibility to reverse time, and then kill one's own granddad prior to the time that one's dad was born. In any case, small number of researchers believe that Catch-22s can be kept in a safe distance through either the Novikov self-consistency rule or the notion of creating parallel universes.

Tourism at the time of:

Stephen Hawking has proposed that the absence of tourists from the future to come could be a defense against the possibility of time travel.

This is a variation on this variation of the Fermi Catch 22. Naturally, this doesn't be a proof how time travel could be physically impossible as it could be the case that it is feasible, yet it never is created or used with any thought regardless of whether it was created, Hawking notes somewhere else that time travel could be possible in a particular spacetime region that has been altered in the correct way, and on the possibility that we cannot create this kind of place in the near future the time travelers would not be able to return to the past prior to this date, and "[t]his image would explain the reason" this world isn't currently invading by "vacationers from the future."[22 This is basically saying that that if an actual time machine were to be built then we won't be able to observe time travelers. Carl Sagan additionally once proposed the possibility that time travelers might exist, but are hiding their presence or not considered to be time travelers. [23]

General relativity

General relativity provides an experimental basis that suggests the possibility of time travel back to the past in some irregular circumstances however, despite the fact that theories of semiclassical gravity suggest that when quantum effects are merged into general relativity, the escape clauses may be closed. This led Hawking to elaborate on the sequence insurance theory in his statement that the most fundamental nature laws prevent time travel, but scientists aren't able to make an informed conclusion regarding the question without a quantum gravity, which would join quantum mechanics with general relativity into an entirely integrated theory. [23][26]:150

Chapter 2: Time Travel To The Past

According to the theory of relativity matter that moves faster than light, starting at one place and then moving onto the next will appear within an inertial case of reference as being in reverse. This is a result of the relativity of synchronization within special relativity. It states that in certain instances, distinctive references differ depending on the possibility that two events at different locations occurred "in the past" or not, and they could also differ on the order of the two instances. In reality, these contradictions occur in the event that the spacetime interval between the two events is "space-like which means that neither of the events occurs behind the one's light cone than the other. [28] If one of the two occasions speaks to the sending of a sign from one area and the second occasion speaks to the gathering of the same sign at another area, then the length of the sign is moving at the velocity of light or slower, the arithmetic of concurrence guarantees that all reference outlines concur that the transmission-occasion happened before the gathering event. [28]

However, because of the possibility of a speculative object moving faster than the speed of light will always be some edges where the body was present prior to when it left, meaning it could be considered to have moved backwards in time. Additionally, as there are two main theories of relativity states scientific laws should to follow the same course for every inertial edge so if it is feasible for signals to reverse time at any edge, it is feasible at every edge. This means that if a witness A transmits a message to a spectator B that changes forward FTL (quicker than light) at the edge of An's, but in reverse time within B's case, and that, B responds with an answer that shifts FTL at B's edge however in reverse within the edge of An's It could be the case that A receives the answer prior to making the initial flag. This is which is an obvious violation of causality at every edge. An illustration of this circumstance using spacetime outlines could be seen here. [29] The scene is described here and there at as a Tachonic telephone.

In accordance with particular relativity it will require an endless quantity of energy in order to accelerate an object that is slower than light up to speed comparable to light(c). Despite the fact that relativity does not deny the possibility of tachyons moving faster than light in all circumstances in the event of a breakdown using quantum field theories and quantum field hypothesis, it seems that it is not be possible to use them to transmit data faster than light. There is also no consensus of the existence of tachyons as the neutrino that is faster than light characteristic had raised the possibility that neutrinos might be tachyons. However the results of the test were found to be untrue following further research.

The general relativity theory extends the speculative hypothesis to include gravity, demonstrating that it can the flow and ebb of spacetime caused by mass-energy and the force stream. General relativity shows the universe in the conditions of field and explanations for the conditions that allow for what's known as "shut curvatures that resemble a time" which, in turn, permit time travel in the past. The first was suggested by Kurt Godel, an answer that is known as the Godel measure, but his (and many other's) theory demands that the universe has physical properties that it does not appear to possess. The question of whether general relativity is able to deny time-like bends under every reasonable scenario is unclear.

Wormholes are an speculative twisted spacetime that is also permitted to be allowed by Einstein field general relativity rules,[31however, it is not possible to travel through a wormhole, unless it was an uninvolved wormhole.

While time travel into the future is possible through at least one of the ways as demonstrated in the previous chapters(time dilation) however, time travel into the past isn't simple and is a problem because of a number of reasons, including there is the Grandfather contradiction, Einstein's particular theorem of relativity and also observations about the absence of future tourists and the obvious observation that absence of matter or ways to exceed speeds of light. There are many solutions to these issues however.

According to Stephen Hawking points, the absence of tourists from the future could be an indication that time travel into the past isn't feasible. This could be resolved by imagining that time isn't a singular the continuum is infinite in "times". So, any person who violates this continuum can teleport their self to any one of possible time variations and have a chance of being one/infinity. In practice, the likelihood that we will observe someone from the future is actually one infinity, which is almost the same as a zero.

It is plausible since it eliminates the Grandfather paradox. With an endless number of "times" and a zero-percent chance of going back to your own time, the probability of killing your grandfather and make possible your existence zero.

It is true that this solves issues related to casuality, it doesn't provide ways to travel into the past, which is to surpass speeds of light. Speed of light that is greater than is not possible because of the way that the present Lorentz transformation is constructed which means that when you attempt to calculate the time dilation of the future using speed $> c$ by using the Lorentz transformation, you'll end up with an no result that isn't certain, and doesn't make mathematical sense, in the sense that it's impossible. There are two possible ways of looking at it either Einstein was mistaken (unlikely) or C cannot be as high, or: C indeed can't be exceeded , but it could be "exceeded" in the event that the multiverse theory is real.

This is another intriguing argument in support of this "time continuum" theory, and the notion the existence of endless possibilities for time and not only one. The fact that there are 'many parallel times' is in complete agreement with the concept of the multiverse as well as the many spaces that exist beyond our planet. In addition to being an instrument for measuring things, but in addition as physical measurement it is logical to observe time with different lengths and with different starting points in the event that there are other universes. So, finding that there are other universes in existence is likely to show that time continuums are not the same and that traveling to new universe will result in different points in time.

If the multiverse theory is true but, how do we travel to other universes caused the possibility of "accident" the possibility of time traveling to the past?

This is a question that's not easy to answer. The first reason is that the multiverse theory is a mere speculation, the first step is finding out about such universes. This becomes more challenging due to the fact that these universes are likely to have characteristics that are different from ours - which means that interaction isn't possible since they're not made of the same material as ours.

The constant expansion that is the world was believed as being caused by gravity-related influence from an alternate universe, but there is no evidence that a similar other universe has been confirmed or even considered. The problem is that, even if it does exist, its unique nature will render detection difficult. For example, dark matter which accounts for the majority of our universe doesn't interact with "our matter", the so called baryonic matter, and the only interaction is via gravitarion -

despite the fact that it accounts for the majority of our universe we're still having trouble detecting it - imagin the case of another universe, they may be invisible/undetectable even if they have impact on our own universe.

What can we do with this universes to travel into the past but? It's firstly likely that the multiverses that we'll eventually be able to identify one time in the future are "tuned" to allow time travel. This means: They're far away, impossible to connect with, and time-symmetric (similar to our concept of time) and current one) ...), and other problems. This leaves us with only one possibility: creating a an entirely new universe! With the present technology, this falls within the realm of science fiction but not as far since the concept is said to be a source of concern for even the most brave scienc fiction authors.

The energy of the visible universe will be 10^{69} Joules or way above that, the total enegy consumed on Earth in 2008 was $5 * 10^{20}$ Joules or $2 * 10^{48}$ or...2000000000000000000000000000000

00000000000000000, this small number whose name is never used in comptemporary science is the amount of times the energy of the observable universe is bigger than the energy we consume here in whole year as billions of people. Additionally, we'll need to master femtotech in full this is the science that will allow us to manipulate the world at the proton scale. It's still early with nanotechnology. After that, we'll need to explore microntech and then eventually get to the atom scale - with the speed of advancements in technology this could take many years, particularly considering how we require a complete understanding of atoms to manipulate them, and remember that an atom is one/billionth the size of a sentimeter. We require technology to transform every atom in the human body into alien matter that will interact with the new universe. It's possible to achieve this in the next few thousand years, but it is definitely not science fiction at the moment.

The process of creating the "time machine" back to the past could appear quite strange and unattainable using the current technology.

1. Create a mini-verse, a new universe with exotic matter developed in a laboratory, the most elemental component ("light") is speedy surpassing the speed of light.

2. The laws of physics within our universe aren't broken, the new universe will remain unobservable and thus interaction with it is impossible, not even through gravity. So you'll need to alter your body, or create an device that allows you to travel to the parallel universe.

3. In the parallel universe you'll be traveling in a direction slower than the speed of light at our local limit, however, it is faster than light speed when measured in our personal universe.

4. Returning to our own universe, and then returning to normal matter, it'll seem like you've not started your journey in the moment and based on the speed of travel and the time that you spend in the newly created mini-verse

Chapter 3: Technology: Current Technology
A proposed time-travel device utilizing the navigable wormhole could (speculatively) operate in the following manner The one part of the wormhole gets speeded up to a crucial division of the speed of light, perhaps using an engine driven drive framework then it is redirected to its beginning the point. However, an alternative is to select one of the passageways in the wormhole and place it inside the gravity field of an object which is heavier than the next passageway before moving it to a point that is near the next one. Both of these methods time-widening causes the wormhole's final stage which has been altered to be necessarily the stationary one, as observed by an outside eyewitness. However, be however the fact that time is synchronized differently with the wormhole, and not outside it, and so the timekeepers that are synchronized at both ends of the wormhole always remain synchronized, as observed by the person who is looking through the wormhole regardless of the way that the two ends change. It is a sign that anyone who enters the speeded end

would exit the stationary finish when it was the same age that the fastened end was prior to section. For instance when, prior to going through the wormhole, the observer observed that a clock at the speeded end read a date in 2007 whereas a one at the other end's clock read 2017 and the person watching would then depart from the stationary end when the clock reads 2007, a move that is reversed in time according to witnesses outside. A major drawback of time machines is to the extent that it is only possible to travel back as far to the time of original manufacturing of the device;[33] in essence, it's more of a method to traverse time than an instrument that travels through time. It cannot allow the technology itself to go backwards in time.

Based on recent speculations about the development of wormholes, the development of a wormhole navigable will require the presence of a substance that has negative energy (frequently called "exotic substance").

More so the wormhole spacetime needs an energy transfer that causes damage to different circumstances [34]. But, it has been discovered that quantum impact can trigger very little or no infringement of the invalid energy. [34] Many scientists believe that the necessary negative energy could actually be feasible because of the Casimir effect, also known as quantum physics. Although the initial calculations indicated that a significant amount of negative energy was required, later calculations showed that the measurement of negative energy could be reduced to a small amount. [36]

in 1993 Matt Visser contended that the two mouths of a circular wormhole having this kind of impelled clock contrast could not be joined without triggering a gravitational and quantum fields that could either cause the wormhole collapse or make the two mouths would repel one another. In the event of this it is impossible for the two mouths to be connected enough to allow for causality infringement. In any event in a paper from 1997, Visser conjectured that a stupendous "Roman Ring" (named in honor of Tom Roman) setup of the number of wormholes in a symmetrical polygon could currently function as a"time machine," in spite of the fact that Visser thinks that this is more likely an error in the traditional quantum gravity theory in contrast to the proof that causality-related can be a possibility. [38]

Different methods in light of general relativity

Another approach is to use a turning chamber more often is mentioned as an Tipler barrel, an arrangement discovered by the late Willem Jacob van Stockum[39] back in 1936 , and Kornel Lanczos[40in 1924, but it was not thought to permit closure timelike curves[41] until an investigation by Frank Tipler in 1974. If the barrel is infinitely long and is able to turn at a sufficient speed around its pivot point and a spaceship flies through the chamber in the winding route could be able to go back to the past (or forward, depending on the direction in which it is winding). However, the size and speed of the process is astonishment to the point where conventional material isn't strong enough to allow it. It is possible that a similar device could be made from a massive string, but none is yet known or are believed to be feasible to create another string of this size.

The physicist Robert Forward noticed that a reckless application of general relativity with quantum mechanics is a better way to create the concept of a time machine. A massive nuclear core in an attractive solid field could expand into a barrel with a dimensions and "twist" can be used to build an actual time machine. Gamma beams projected at it could allow data (not be it any different) to be transmitted back in time. Regardless the fact that he pointed out that , until we can come up with one hypothesis that connects quantum mechanics and relativity we'll have no idea whether these theories are true or not.

Another major complaint about time travel plans , in light of the possibility of turning barrels, or incalculable strings was presented through Stephen Hawking, who demonstrated the hypothesis the general relativity theory that it's difficult to construct the time machine that isn't common (a "time machine with a minimalistically constructed Cauchy cityscape") within a region in which the condition of energy that is feeble is satisfied, meaning that the region is comprised of having a zero energy (exotic material). The most common arrangements are Tipler's accept large barrels lengths, which are easier to decompose scientifically and despite their size, Tipler advised that a smaller chamber could produce closed time-like bends when the pivot speed was sufficient, he failed to show this.

In any event, Hawking calls attention to the fact that, as a result of his assertion that "it isn't feasible with the thickness of positive energy all over! I'm able to show that in order in order to construct a limited-time machine, you need negative energy."[26]:96 This result is derived from Hawking's 1992 paper about the order assurance guess in which he analyzed "the possibility that causality infringement is visible within a narrow area of spacetime that does not have ebb and flows singularities" and proves that "[t]here is an Cauchy skyline that is minimally constructed and when all is said and done , contains some or all closed invalid geodesics that are deficient. It is possible to identify geometrical quantities that quantify what is the Lorentz support and the region's increment upon going around these invalid geodesics. If the causality infringement is derived by a non-compact start surface, it is the middle of the frail energy state is a violation of"the Cauchy horizontal horizon."

But this theory does not discredit the possibility that time travel could be possible (1) via time machines that use non-minimalistically-produced Cauchy skylines (for instance the Deutsch-Politzer-designed time machine) and (2) in areas that contain mysterious matter (which could be necessary for navigable wormholes, or an Alcubierre drive). Because the theory is based upon general relativity it's possible to develop a hypothesis for quantum gravity that would replace general relativity would allow time travel, even with no interesting subject matter (however it is also possible this hypothesis could put restrictions on time travel or even completely ban it, according to Hawking's order certainty conjecture).

Certain tests conducted give the impression of a reversed causality but they are susceptible to provide a better understanding. For example in the postponed decision quantum eraser test carried out by Marlan Scully, the matches of snared photons are divided in "sign photons" and "idler photons" and signs photons rising from one of two regions, and their positions being later determined as in the twofold test. Depending on the way the idler photon will be classified, the researcher will be able to determine what area the sign photon erupted from or "delete" the information.

Despite the fact that photons that signify the obstruction are able to be measured prior to the decision is made about idler photons the choice appears to be made retroactively, determining whether or not an obstruction instance is watched when connecting estimates of idler photons with the signal photons that are comparing.

Even so, because obstruction has to be detected after photons that are idle are taken into account and are linked to the sign photons there is no way the experimenters to know which decision will be taken in advance, just by looking at the sign photons. However, under the most quantum mechanics-related translations the results can be explained in a manner that doesn't harm causality.

The experiments of Lijun Wang could also indicate causality violation because it was possible to move waves through the caesium gas knob in a manner where the bundle appeared to exit the globule just within 62 nanoseconds of its entry. In reality it is true that a wave bundle not a singular, highly known issue but rather an array of different waves with different frequency (see Fourier exam) The bundle may appear to move faster than light, or even in reverse in the course of time, regardless of the fact that any of the unadulterated waves generally behave in this way. The effect isn't able to send anything information, energy, or even data faster than light. this experiment isn't intended to harm causality.

The scientists Gunter Nimtz as well as Alfons Stahlhofen who are from The University of

Koblenz, case that Einstein's idea was weakened of relativity by sending photons faster than the speed of light. They claim to have supervised an experiment where microwave photons traveled "immediately" between two pairs of crystals which had been elevated to 3 feet (0.91 meters) separated using the technique called quantum burrowing. Nimtz stated in New Scientist magazine: "Until further notice this is the largest incident of exceptional relativity I have observed."

Shengwang Du has claimed in a follow-up diary that he had watched single photons' precursors and claimed that they are no faster than the speed of c in vacuum. The experiment involved moderate light as well as moving light through the vacuum. He made two single photons, one of which passed through rubidium molecules which were warmed by lasers (subsequently altering the light) and the other through an air vacuum. Both times, it is clear that those that preceded them were the principal bodies of the photons and the antecedent was through c in a vacuum. According to Du this implies that it is not possible to prove light traveling at a speed faster than the speed of c (and thus, not recognizing causality). Certain people from the media interpreted this as proof that time travel into the past by using superluminal velocities could not be achieved. [46][47]

Experiments that do not involve science:

Krononauts:

Certain studies were conducted to entice future generations who could create technology for time travel to come back and show it to the people who are in the present. For instance, the Perth's Destination Day (2005) or the MIT's Time Traveler Convention vigorously promoted permanent "commercials" that offered a suitable meeting location and time where future travelers could gather. The year 1982 saw a group at Baltimore, Maryland, distinguishing itself as the Krononauts organized an event that invited guests from the future. The trials had the potential of delivering positive results that demonstrate the existence of time travel. However, they the trials have failed as a result--no time travelers are reported to have attended either of the events. It is possible that future generations have traveled back to the past, but have not ventured back to the time of the meeting and location in a different universe. [51]

Another factor is that for all of the devices for time travel that are utilized in current technology, (for example, those that use various wormholes) it's difficult to travel back prior to when the time machine was actually invented. [52][53]

Chapter 4: Time Travel To The Future:

There are many ways that a person could "go in the direction of the future" in a restricted sense. The individual can set up things so that, within their small amount of subjective time the majority of their subjective time is gone for people in Earth. In this case, for instance, a viewer could take a trip off the Earth and return at an extremely fast speed, the trek lasting just some time in accordance with the eyewitness's own clocks, and then return to discover that a significant amount of time has passed in the world of Earth. According to relativity, there's no definitive answer on the question of what amount of time that "truly" occurred during the trek. It could be equally significant to claim that the trip continued for just two years or that the trek continued for many years dependent on the selection of the reference edge.

This kind of "go into the future" is seen on tiny time scales, and is applicable to any time scale. It is possible to connect it using speed-based time expansion in line with the relativity hypothesis for instance by travelling around the speed of light to a dispersed star, and then reversing before pivoting, then moving almost at the speed of light and returning to Earth as described within the dual mystery. It is possible to connect it using gravitational time expansion under the theory of general relativity that is, for instance, by dwelling in an uninhabited, massive and dense matter, or by living away from the space of a black hole or even living as an explorer whose weight and thickness allows for adequate gravitational time expansion around to it. [54][27]:33-130

Time dilation:

Time expansion is permitted by Albert Einstein's particular and general theories of relativity. These theories state that with respect to an observer, time moves in a more gradual manner for bodies that move rapidly with relationship to that observer or those which are more deep in a gravity-based plane. [55]

For instance, a timer that is moving with respect to the observer will be rated as running moderately within the rest of the viewer's outline. As the clock speeds up at the speed of light, it slows down to a halt even though it will never reach speed at which it could not completely stop. If two timekeepers are moving inertially (not increasing speed) relative to one to each other, the effect is equally each clock observing the other clock to be slower. But this symmetry breaks when one clock speeds up, such as is the case in the two-way Catch 22 where one twin remains focused, while the other is in space and pivots (which involves the speed of its ticking) and then comes back. In the two clocks, both agree that the twin that is traveling has a lower rate of maturation. General relativity asserts that time dilation effects also occur on the chance that one check is dependent on an gravity well as opposed to another, and the clock in the well ticking more slowly. This effect is to be considered when aligning the timekeepers of each satellite of the Global Positioning System, and it can cause huge differences in maturing rates for people who are at different distances.

It was estimated that, in general relativity, an individual could move forward at four times faster than inaccessible observers in the form of a circle that has a diameter five meters in size and as large as Jupiter. For such a person who is constantly adjusting his "own" time, they would have a time equivalent to four seconds away witnesses.

There's a wealth of evidence to support the validity of the conditions for speed-based time widening in relativity[56] as well as gravitational time dilation. The most well-known and easily reproducible example is the perception of muon decay in the air. With the advancements in technology, it is possible to allow an explorer of human race to not age as fast than the other humans are on Earth by a fraction of a second. The current record is around 20 milliseconds by the Cosmonaut Sergei Avdeyev.

Time dilation can be clearly speeded up for living creatures by hibernation, in which metabolism and body temperatures of animals decreases. Another compelling way to accomplish this is to suspend movement that allows the rates of compound processes within the subject would be drastically reduced.

Time expansion and suspended living only permit "going" into the future, not the past, therefore they do not affect causality and they are an open question when it comes to being described as time travel or not. But time dilation could be considered to be a better one to understand the concept of "time travel" rather than suspended activities, since as time expands, the time that passes by is shorter for the person who is explorer than those who remain behind. Therefore, the person who is explorer is believed to have reached the future earlier than others even though in suspended activity that is not the case.

Different ideas from the normal physical physics

Parallel universes might provide an opportunity to escape the trap of Catch-22s. Everett's multiple universes clarification (MWI) that is based on quantum mechanics suggests that quantum events of all kinds may occur in totally separate timelines. These different histories, or parallel ones, could form a stretching tree that represents every possible outcome from any communications. If every possible outcome is present the possibility of any Catch-22s can be resolved by introducing unfathomable events occurring inside an alternate reality. This concept is often used in science fiction, but certain physicists for instance David Deutsch have proposed that the possibility of time travel exists in the event that the MWI is correct that a time-traveler is likely to end into a different time in a different place than the one that he came from. However, on contrary, Stephen Hawking has contended that, regardless of the likelihood it is true that MWI is true it is essential to assume that all time travelers will experience a single, predictable timeline, so that time travelers remain in their own personal universe instead of traveling into an alternate reality. The scientist Allen Everett contended that Deutsch's method "includes altering the fundamental quantum mechanics standards and goes beyond simply embracing quantum mechanics and the MWI". Everett also states that, regardless of the likelihood that Deutsch's method is correct it suggests that any natural-looking traveler composed of multiple Pframe references could be of a distinct grouping when travelling back through time via the wormhole, and with different Pframe references emerging in different universes. [67]

Daniel Greenberger and Karl Svozil proposed that quantum hypothesis provides an explanation for the concept of time travel without contradictions. The quantum hypothesis causes conceivable state to "crumple" into one state. Thus the past seen in the present determined (it has a distinct possible state) however, the present that we see from the past contains a variety of possible states, until our actions make it fall to one single state. Our actions will be seen as inexplicably.

Quantum physics:

Quantum-mechanical marvels such as quantum teleportation or quantum teleportation EPR , also known as quantum ensnarement could appear to be a part which takes into consideration faster than light (FTL) communication or travel time, or actually, a few quantum mechanics translations such as the Bohm explanation presumes that data is transferred between frame references at a time taking into consideration the final aim of keeping the relationships between frames. It was described as "spooky activities at a time of separation" according to Einstein.

The method of ensuring causality by quantum mechanics a comprehensive result of current quantum field hypotheses and therefore, advanced theories do not consider the notion of time travel, or FTL correspondence. In the case of a specific instance in which FTL was proposed in more specific studies, it has proven that in order to obtain an indication, some form of conventional correspondence must also be employed. The hypothesis of no correspondence is also a general affirmation that quantum snare cannot be used to send data faster than conventional signals. The fact that quantum phenomena don't enable FTL timing travel, is usually neglected in the general scope of quantum teleportation studies. What quantum mechanics are used to preserve causality is an ongoing area of study.

Chapter 5: Philosophical Understandings Of Time Travel:

Time appears to follow an agenda, to be naturally directional. The past is behind us and remains unaltered and unchanged and is accessible through memories or written documents The future, is ahead of us and isn't always settled and, despite possibility that we could see it in some way but we don't have any concrete proof or evidence. A majority of situations we experience are irreparable such as, for instance it's easy breaking an egg and it is difficult, if not impossible, to break an egg that is officially broken. It's hard to imagine that this particular movement could be in another direction. This restricted direction or asymmetry of time is frequently referred to by its name, the "arrow" of time and it's what that provides us with an impression of the time passing, our progressing through various minutes. The arrow of the time that is, then, the unchanging and impressive bearing that is connected to the obvious and inexplicably "stream that is time" into the future.

The possibility of an arrow of time was first investigated and developed to varying degrees by the British scientist and space expert Sir Arthur Eddington in 1927, and the origin of the concept is generally believed to be his. What attracted Eddington was the fact that the same arrow of the time could be applicable to an alien race from the opposite part of our universe that is the case for us. This is and has nothing to do with neuroscience or brain research but with how the universe functions. The timeline isn't unrecognizable from the time itself, but instead an indication of the universe's nature and how it has evolved.

Does The Arrow of Time an Illusion?

In the section of Relativistic Time, as indicated by the Theory of Relativity, the reality of the universe can be represented by four-dimensional space-time with the aim of proving that time is not a "stream" however it just "is". The appearance of an the arrow of time we experience in our daily lives as a result of this gives the impression that it is a dream in this version of the universe. It is an increasing quality is a consequence of our unique presence in the development in the evolution of our universe.

Perhaps most remarkable and intriguing is the fact that, even though there are events and processes at the visible level - like the behavior of the mass of materials that are part of our daily life are unmistakably wrong in time (i.e. normal procedures are).

are characterized by a brief request with evidently a forward-looking course of time) physical processes and laws at the sub-atomic scale, whether they are established, quantum or relativistic or quantum, are all at least partially time-symmetric. In the event that an physical procedure happens to

If it is physically possible, then in general, so is an identical procedure that is run in reverse, meaning that even if you were to hypothetically observe a film that demonstrates the physical process and you were to watch it, you would not be able to discern when it was being played in the forward direction or reversed, since both are equally plausible.

In the general way, most of laws of science do not really define a arrow of time. There is in fact, be that as it might, a crucial special instance that is that of the Second Law of Thermodynamics.

Thermodynamic Arrow of Time

The majority of the observed transient asymmetry at the level of perceptual asymmetry - the reason we think the time period to be in an upward direction - ultimately boils down to thermodynamics, which is the study of the effects of warmth and its relation to working or mechanical energy and later, some specifically relate to that of the Second Law of Thermodynamics. This law states that when one advances in time and time, the net the entropy (level of confusion) of any closed or disengaged framework will always increase (or at best, keep the momentum).

The concept of entropy and the decay of frameworks that were requested was studied and clarified by the German physicist Ludwig Bolzmann in the late 1870s, building upon the earlier ideas from Rudolf Clausius, yet it is a thorny and frequently misinterpreted thought. Entropy is often interpreted generally as implying that objects (matter or energy, and the like) tend to disperse. This is why hot questions always release heat to the air, and then chills out, but it's not the other way around espresso and drain are two of the most common however they aren't separate; a house abandoned without a guard will eventually crumble away however a pile of blocks will never be able to frame itself up into a building or a house.

As we've discussed earlier the subject isn't as simple as that and a higher mental state could be as an urge toward arbitraryness.

It is important to note that in thermodynamic structures which aren't closed there is a possibility that entropy will decrease with the passage of time (e.g. the arrangement of particular precious stones, the various living structures that could reduce nearby entropy to the disadvantage of the environment around it and result in an overall rise in entropy. The formation of segregated pockets gases and tidy in stars planets have planets, despite the fact that the global entropy is constantly growing and on). The limited or temporary examples of the universe's arrangement are in the manner of epiphenomena in the general picture of a universe that is moving rapidly toward scatter.

It's also possible to be strange, however it is all-in-all true it is true that entropy in general grows even in huge forms of structure are observed within our universe (e.g. cosmic groups, systems or fibers and other such things) and that the thick and narrow dark openings have extremely high entropy and represent the brain's lion's share of the entropy that exists in the present universe. The moderately smooth structure of the universe's early years (see the section about Time as well as The Big Bang) is really an indication of low and high the amount of entropy (i.e. the high level of entropy doesn't necessarily indicate smoothness; an irregular "unevenness" similar to that of the present universe, is actually the norm for high levels of entropy).

The majority of the processes that seem to be inexplicably irreversible over time are ones that begin due to reasons that are not clear with an extremely unique highly sought-after state. A deck of cards is in demand for numbers, but when we arrange them, they become disorganized and an egg is extremely sought-after state than breaking or baked egg and so on. There is no law of science that prevents the practice of arranging an entire set of decks from resulting in an amazingly requested arrangement of the cards There is always some chance of it happening however it's only the most unlikely possibility. In a different way that there are many possibilities of scattered plans in the jigsaw than there is single requested plan that creates an entire picture. This is why the apparent asymmetry of time is actually an illusion of chance. things evolve from request to chaos not due to the fact that the fact that turning around is a mystery however, but because due to the fact it's extremely difficult.

The Second

Law of Thermodynamics is along this same line, more of a measurable principle than a fundamental law (this was the genius of Arrowzmann). In any event, the conclusion is that, if the fundamental state of a framework is one with moderately high demand and the tendency will be essentially always towards problem.

Thermodynamics appears to be, from all reports, to be among the most fundamental physical processes that is not time-symmetric consequently, so fundamental and ubiquitous is it throughout the universe, and it may be without aid in our perception of time as being a course.

Sure it is true that a few of the different arrows of time seen below (apparently) eventually come back to the symmetry of thermodynamics.

It is true that this law is so precise it is clear that estimation of the entropy has developed as a method of recognizing what was past and what is future and the thermodynamic arrow of time has been suggested as a reason why that we are able to recall the past but not anticipate the future because of the fact that the amount of confusion or entropy was less earlier than later.

Cosmological Arrow of Time

It has been suggested it is believed that the arrow that defines time is directed towards the evolution of the universe because the universe has been expanding and growing since the Big Bang (see the area on Time and the Big Bang). It became apparent at the beginning of the 20th Century because of the research from Edwin Hubble and others, that space is no doubt expanding, and that cosmic systems are

The constant motion of the universe can cause separation. In this manner, it is evident that it appears that at a pre-existing time the universe was smaller, and then it was a single singularity or point that is known as"the Big Bang. The universe therefore appears to be a particular (outward) orientation. Our day-to- life regardless we are not aware of this fact and it's difficult to

We can perceive the expanding universe as an the arrow of time.

The cosmic arrow of time may be related to, or even subordinate to the thermodynamic arrow. This is due to the fact that the universe continues to expand and moves toward the extremist "Warmth death" also known as "Enormous Chill" It is also going through a process of increasing entropy until reaching a point with the greatest entropy. This is at which point the quantity of energy usable is likely to be non-important, or even insignificant.

This is in line in accordance with this being in line with the Second Law of Thermodynamics in that the overall direction of the universe is from the current semi-requested statethat is separated by outcrops of request and the structure, and towards a completely disjointed state with warm harmony. One of the most interesting mysteries in the present-day science, but is the reason the universe was characterized by a lower entropy level at the time of its creation as well as in the Big Bang.

The cosmic arrow of time could be linked with, and possibly dependent on the thermodynamic arrow. Considering that the universe continues to expand and moves toward the extremely "Warmth death" also known as "Enormous Chill" the universe is going through a process of increasing entropy and eventually reaching a point that is the most extreme in entropy in which the amount of energy that can be used becomes to be irrelevant or even non-existent.

This is in accordance in accordance with Second Law of Thermodynamics in that the general heading comes taken from the current semi-requested condition separated by

Outcrops of request and structures and structure, leading to a completely confusing state of the warm balance. What is still a mystery to scientists of today, yet, is exactly why the universe experienced an entropy that was low in the first place which was it was the Big Bang.

It is also possible although it is more remote in the light of the forecasts of the present research - that the current expansion time of the universe might be extended to a longer period of time, then slow or stop and then afterwards, reverse itself under the influence of gravity. The universe could then shrink and return to a reflection that reflects what was the Big Bang known as the "Enormous Crunch" (and maybe an eventual "Huge Bounce" as part of a series of repeating cycles). When the universe expands and shrinks, entropy will generally decrease and, presumably the arrow of the time will begin to turn and time will eventually begin to reverse itself. In this scenario the arrow of time we are experiencing is only a part of our current position in the evolution of the universe. And in the future the time, it might change direction. However there are some nebulous aspects to this perspective in the sense that, if looking at it from a somewhat distant and long-term perspective, time is always moving forward "advances" (in certain areas at least) regardless of whether the universe is in a withdraw stage instead of the extension phase. This is why the infinity asymmetry of time can be observed at the moment even in a "shut" universe like this one.

A Radiative Arrow for Time

The waves, like radio waves, light waves water waves, sound waves, etc are always radiative and radiate outwards from their source. While theoretical conditions consider the opposite (covergent) waves but this isn't observed in the natural world. This asymmetry is thought of by some as a reason to explain the asymmetry of time.

It is possible that the arrow radiative could also be linked to the thermodynamic-based arrow since radiation is believed to promote increased Entropy, while meeting suggests an the expansion of request. This is particularly evident when we think of radiation from a molecule perspective (i.e. consisting of photons) as quantum mechanics suggests.

Quantum Arrow of Time

The whole quantum mechanics (or in the absence of anything else, the conventional Copenhagen concept of it) relies on Schrodinger's Condition along with the crumple of the wave capabilities (see the section that is based on Quantum Time), and it appears, from all indications to be a time deviation marvel. In particular, the surface of a molecule can be represented through a wave capacity which is essentially a probability scale that gives various probabilities that the molecule could be in various possible locations (or superpositions) as well as the wave work is merely crumpled when the molecule actually observed. The molecular can be considered to be in a specific position, and the information from the wave capacity disappears and isn't able to be replicated. In this way this process is irreversible and time-independent and an timeline is created.

Some physicists, such as the group comprised of Aharonov, Bergmann and Lebowitz in the 1960s, have studied the findings, but. The tests they conducted assumed that we can only find unbalanced time quantum mechanics solutions when we ask time-topsy-turvy-related questions as well as that research and tests can be restricted in a way that results are time-symmetric. In this regard quantum mechanics doesn't create a time asymmetry in the world, but rather the world forces time asymmetry upon quantum mechanics.

It isn't certain what the quantum arrow of time, even if there is no doubt it exists at all by any stretch of imagination, is connected to different arrows. However, it is possible that it's linked with the thermodynamic arrow which has a tendency to cave in wave capacities to higher entropy states , and bringing lower ones.

Frail Nuclear Force Arrow of Time

Of the four major forces in science (gravity electromagnetism, gravity, the solid atomic drive, and the atomic constraint that is powerless) The powerless nuclear drive is the one that only does not usually exhibit finish time symmetry. To a certain extent it is believed that there is a weak the compel arrow, and it is the primary time arrow that creates the impression of not being a significant part of that of the thermodynamic arrow.

The insignificant atomic compel is a remarkably ineffective cooperation at the center of a molecule. It is responsible for among other aspects, radioactive beta rot and the production of neutrinos. It's possibly the least understood and the most bizarre of the major advantages. In certain circumstances, the time-reversible constraint is a problem, e.g. protons and electrons can be crushed to produce an electron and a neutron or a neutron, and the neutron and the neutrino collide CAN be able to deliver an electron and a proton (regardless of the fact that the chances of this happening are very low). In any event there are instances of unreliable collaborations that is irreversible in time like the example of the swaying, rot and tilt of neutral kaon, and hostile to particles of kaon. If certain conditions are met,

It has been proven in a tentative way that kaons and Kaons actually rot at different rates, which shows that the compel without power is

Not necessarily time-reversible which is why it creates a type of an arrow of time. It is important to note that this is not an essential or solid time-based arrow like it is a thermodynamic arrow (the distinction).

is a process that can go in any direction but in a slightly remarkable way or at a different speed, and an irreversible control - similar to the entropy phenomenon - which can't be in any manner, manner, or form be reversed). In fact, it's an extremely rare event which is so small and barely visible in its effects and thus adrift from other arrows that it is usually presented as a bizarre anomaly.

Causal Arrow of Time

Despite the fact that causality is not directly associated with the sciences, causality appears according to all indications to be a personal tie by the arrow of time. In the definition of causality the cause is prior to its effect. Even though it's a challenge to be able to accurately define circumstances and their outcomes however, the concept is evident in our daily life. In the event that you drop a wine glass on a floor that is hard, it's likely to break. However, smashing glass in the floor is not likely result in a broken wine glass. When we cause something to happen in some way, we are to a certain degree influencing the future, however, regardless of what we do, we are unable to alter or influence the past.

The end of the day however, the basic principle could be to be to return in the direction of the thermodynamic bow While smashed glass scattered around can easily be created out of a much demanded wineglass, the flip around can be a lot more difficult and is not possible.

The Mental Arrow of Time

A variant that is a variation of the causal arrow can be sometimes referred to as the perceptual or mental time arrow. We be born with a awareness that our perception is moving from the past to the elusive future. We anticipate the unknown and then naturally work towards it. And when we do remember about the past but we do not often waste time trying to alter the present-day and changed past.

Stephen Hawking has contended that even the mental timeline is now reliant upon the thermodynamic arrow and that we could remember the past as they form a comparatively small set in contrast to the endless number of possibilities for future sets.

Human-centered Principle

Some masterminds, like Stephen Hawking once more, have remained on the arrow of time what was time called the human-centered rule, the idea to believe that science's laws work solely because of the fact that these rules allow the

the development of conscious, addressing species like us. There is no way that the universe is "outlined" to allow humans to be human, but that we are in the same universe because it is what it appears to be even though the universe could not without much of a stretch have been created in a unique way with completely distinct laws.

This way, Hawking contends, a strong thermodynamic arrow of time is a prerequisite for the enthralling life we're likely to produce.

For instance, animals similar to us must eat the food (a relatively sought-after type of energy) and transform it into heat (a typically disorganized type of energy) and an arrow that is thermodynamic like that we see around us is crucial. If the universe was any other way, we'd not be around to observe it.

Time travel theories are packed with questions on causality and Catch-22s. Contrary to other important concepts in the field of cutting-edge science, time has not seen with a high degree of clarity. Philosophers have been speculating about the ways in which time works since before the time of the ancient Greek savants. Some physicists and savants who consider the nature of time also contemplate the possibility that time travel could have constant implications. The possibilities of Catch-22s and the possible arrangement of them are usually thought of.

For more information on philosophical considerations regarding time travel refer to on the works of David Lewis. For additional information on scientific theories of time travel, look into research by Kurt Godel (particularly his guessed universe) and Lawrence Sklar.

Presentism versus eternalism

Many experts have suggested that relativity implies eternalism, the possibility that events in a lengthy time are real and not only as modifications that took place or appear in the future. Scientist and philosopher Dean Rickles can't help contradicting some of his theories, but he has he has observed that "the consensus between logicians is on all indications that singular and general relativity don't agree with the current view of. "[73Certain rationalists believe that time is similar to the measurements of space as well as that future events are "as of right now" and in the exact same way better locations exist as well as that there isn't a target stream of time. However the fact that this view is not widely accepted. [74]

The ring and the bar Catch 22 is a case of the relative synchronization. Both ends of the bar go through the ring the same time on the remainder of the ring (left) However, the closings of the bar move with a constant speed through the remaining part of the bar (right).

Presentism is a theory that believes that the future and past are just changes that occurred or will be revealed in the future and have no real existence that is their own. According to this view the notion of time travel is unimaginable due to the fact that there isn't a time or space that one can travel back to. [72The authors Keller as well as Nelson have argued they believe that, regardless of whether future and past frame references are not possible and are not present, there may in all likelihood be different truths regarding the past and future which is why it's possible that a future truth regarding a time traveler's choice to travel back to the present time might clarify the real appearance of the time traveler in the present. theories are being challenged by a number of writers. [76]

Presentism in the established spacetime values that only the present is present and is incompatible with relativity. It was shown in the corresponding case: Alice as well as Bob are both synchronous observers of the event O. For Alice the event E is synchronous with O however for Bob the event E is either in the future or past. This is why Alice and Bob are not in agreement about what happens in the present moment, which discredits the traditional notion of presentism. "Here-now presentism" seeks to make room for this by simply being aware of the time and space of one point. this is unacceptable because frames of reference that travel all the direction to the "here-now" substitute between real and untrue, despite that there is no popular "here-now" that is an example of the "genuine" current. "Relativized presentism" acknowledges that there are a myriad of frames of reference. Each one of which is an alternative arrangement of concurrent events that makes it hard to discern a single "genuine" present. And from now on, it is possible that all instances in the present are real--obscuring the distinction between eternalism and presentism, or every frame of reference is its own unique actuality. The alternatives to presentism in relativity appear, according to all indications to be drained however Godel and others believe that presentism could be a significant factor for certain varieties that are considered general relativity. [77]

Chapter 6: Paradoxes Of Time Travel

The granddad paradox:

A topic that is often discussed in the philosophical debate about time is how to get there, in the off chance that someone could reverse time, Catch-22s might be the result when the time traveler decides to alter the course of events. The most effective example of this is the granddad dilemma and potential autoinfanticide. This granddad Catch 22 can be described as a hypothetical scenario in which a time traveler reverses time and tries to kill his father's grandfather in one go prior to the time his granddad meets his grandma. In the event that the time traveler did this the father of his son was never born or even the time traveler and in that case, the time traveler would never have attempted a reverse that would have killed his grandfather. The problem is sometimes addressed as autoinfanticide, in which an adventurer retreats, and tries to commit suicide as the infant child. If he were to perform such a thing then he would not have become reverse the process and commit suicide as a young child.

This dialogue is crucial to the logic behind time travel, as scholars debate whether these issues render the idea unattainable. Some scholars respond to the Catch-22s through anger that the facts could support the idea that regressive travel is possible, but it's hard to alter time in any way. an idea similar to the proposed Novikov self-consistency norm in the field of science.

Ontological paradox:

It is the Novikov self-consistency principle, named for Igor Dmitrievich Novikov, states that all actions, whether that are made by a person who is a time-traveler or by a person who is able to travel back in time were part of the history from the beginning in this way it is impossible for a individual who is traveling through the time "change" the course of history in any way. The actions of a time traveler could cause events that occurred in their own history however, which raises the possibility of a roundabout causality which is sometimes referred to as an ontological paradox, also known as the bootstap paradox. The term "bootstap" Catch 22 was promoted by Robert A.

Heinlein's novel "By his Bootstraps". [82 The Novikov self-consistency guidelines recommend that the local rules of science within an area of spacetime that is populated by time travelers cannot be less than the local laws of science in a different area of spacetime. [83]

It is the philosopher Kelley L. Ross contends in "Time Travel Paradoxes"[84] that in an ontologically odd scenario that includes a physical object it is possible to have an infraction on the law known as thermodynamics' second law. Ross employs Somewhere as a piece of Time for instance , in which Jane Seymour's character presents Christopher Reeve's character an item of time she's been claiming for a long time and then, when he returns to the past, he presents the identical watch to Jane Seymour's character, who is 60 years older than. According to Ross says:

The watch is an unintelligible product. It ignores what is known as the Second Law of Thermodynamics, the Law of Entropy. In the event that time travel makes this watch feasible, then the idea of time travel is not even conceivable. The watch, however will be completely indistinguishable to itself during the nineteenth and twentieth centuries as Reeve takes it from the future and quickly into the past and gives it to Seymour. The watch, however cannot be indistinguishable from it, as each of the years that it's in the possession of Seymour and, after that Reeve it wears according to the normal way. Its entropy will rise. The watch that is returned by Reeve will wear more than the one which was obtained from Seymour.

But this second thermodynamic law has been considered by the current physicists as a factual law rather than an absolute one, therefore inversions that are unconstrained of entropy or the inability to increase entropy not a new concept, but they are not certain (see for example the variation hypothesis). In addition The second thermodynamic law declares that entropy must increase in frameworks that are isolated from interactions with the external world. Igor Novikov (maker of the Novikov self-consistency rule) has argued that due to natural frame references, such as the watch's worldlines that form closed circles, the external world could make use of energy to repair wear and entropy that the product acquires throughout the duration of its existence thus it will return to its original condition after closing the loop. [27]:23

Hypothesis of Compossibility:

David Lewis' examination of the possibility of compossibility and the consequences of altering the past is designed to depict the possible results of time travel within the form of a single-dimensional creation of time without constructing intelligent puzzles. Take a look at Lewis's situation with Tim. Tim is a snob about his granddad, and may want to kill him. The biggest problem to Tim has to do with the fact that grandfather died a few years ago. Tim is so determined to murder his granddad that he constructs the time machine to travel back to 1955 when his granddad was young and then kill him. In the hopes that Tim will be able to travel back to a time where his granddad still lives and is still alive, the question should be asked: could Tim kill his grandfather?

For Lewis the answer is regarding the usage in the usage of "can". Lewis explicitly states "can" should be viewed in relation to the aforementioned facts that are relevant to the particular situation. Consider that Tim owns a gun and has spent years preparing for a rifle and a straight shot in the morning of a clear day and there is no external power restrict Tim's trigger fingers. Would Tim kill his granddad? Based on these facts, there is you can be sure that Tim may in fact kill his grandfather. In the end, at all times, it is clear that the largest portion of the facts are a part of Tim killing his grandfather. If we are considering the possibility of guilt in any given scenario it is best to put together the most complete collection of possible facts that we can.

Think about the possibility Tim's world is based on the fact that his granddad actually died in 1993, and not 1955. The new information regarding Tim's situation reveals that the fact that he killed his granddad isn't compatible in the current arrangement of facts. Tim cannot execute his grandfather because his granddad died in 1993, but not when Tim was young. In this way, Lewis closes with making the declarations "Tim does not have the ability to execute his granddad because he's got the skills" and "Tim isn't, and won't due to the fact that it's difficult to alter the past" They aren't contradictory and are both valid in the context of facts. The usage for "can" is ambiguous as it is unclear if he "can" as well as "can cannot" according to various relevant facts.

What will happen to Tim while he concentrates on? Lewis is confident that his gun will be able to hold, or the feathered creature could be able to fly through the air or Tim is essentially slipping on the banana peel. In all cases there will be a coherence in the universe that can stop Tim every time from taking out his grandfather. [85]

Pradoxes discussed in sci-fi:

Time travel themes in science fiction and the media can generally be divided into three main categories: the unchanging course of events; a timetable that is variable as well as exchange history (as in the many universes of interpretation). In many stories the term "timetable" is used in order to refer to every physical event that happens, which means that in stories about time travel where the events are rearranged and the character of the time traveler is depicted as creating a different or modified timeline. This usage is indistinguishable from the use of the term "timetable" to refer to a kind of graph which is a particular pattern of events. this concept is distinct from a world line one of Einstein's theories of relativity, which refers to the entire story of a single traveler.

One of the complaints that are frequently made against the concept for time machine technology in science fiction is that they ignore the motion in Earth. Earth between the time the time machine is pulled back from the time it comes back. The possibility that a person is able to enter a time machine that takes them to 1865, and then venture into the same place on Earth could be said to ignore the fact that Earth is moving through space around the Sun that is moving within the system of the cosmos, etc and so those who advocate this theory believe the possibility that "sensibly" that the machine will be able to return in space that is far from the Earth's position on the time of departure. However, the idea of relativity snuffs out the notion of absolute time and space. In relativity, there is no known facts about the space-time separation between events that happen at different times[91for example) (for instance, an event that occurs on Earth in the present and an event that occurred on Earth during 1865) In this way there is no reality about which location in space at the time is in the "same location" as is where the Earth was at another date. According to the theory of relativity, which handles situations in which gravity is non-existent sciences' laws follow along the same route in every inertial reference point and in this regard, each frame's viewpoint is superior physically to any other frame's and distinct edges differ on the possibility that two events at different dates occurred in the "same place" or "diverse locations".

In the concept of general relativity, which combines gravity's effects each direction framework is equal with respect to the element that is called "diffeomorphism of invariance". [92]

Chapter 7: Faster Than Light

Faster than light (likewise Superluminal, also known as FTL) travel and correspondence allude to the emergence of information or matter faster than the speed of light. Based on the bizarre theory of relativity that a molecules (that has a rest mass) that has subluminal speed demands infinite energy to speed up to light speed however, the unique relativity theory does not exclude the existence of pframe references that move faster than light in all instances (tachyons).

But, what handful of physicists describe as "evident" or "viable" FTL[1][2][3][4] is based on the notion that strangely mutilated regions of spacetime could permit matter to reach removed areas faster than light can in conventional or normal spacetime. As evidenced by the current research

Hypotheses about matter are still needed to make a subluminal discovery about the spacetime district that is privately altered and the obvious FTL cannot be ruled out through general relativity. One example of a clear FTL assertions are the Alcubierre Drive and the safe wormhole, despite fact that their physical credibility is doubtful.

Regarding the subject matter of this post, FTL is the transmission of information or matter faster than c. It's a congruous comparable to the speed of light within a vacuum which is 299,792,458 miles per second (by definition) or approximately 186,282.4 miles per second. It isn't the same thing as travelling faster than light.

Certain procedures produce faster than c, but they aren't able to transmit information (see examples in the area of rapidly following).

Light is accelerated at the rate of C/N when it's there is no vacuum, but instead passing through a medium that has a refractive index = n (bringing the issue of reflection) as well as in certain materials, the Pframe reference points can travel faster than the c/n ratio (yet in the same way slower than the c) which can trigger Cherenkov radiation (see the speed of the stage below).

None of these wonders impedes relativity or causes problems with causality as such, and neither of them is FTL in the manner depicted in this video.

In the accompanying illustration Certain impacts could appear to be traveling faster than light, but they do not transfer information or energy as fast as light, and therefore aren't ignoring the fact that they have unique relativity.

Day by day sky motion:

For an Earthbound observer Sky objects undergo a complete upheaval around Earth within a single day. Proxima Centauri, which is the nearest star in the planetary group in the vicinity and is located around 4 light years away. From the perspective of a geostationary view, Proxima Centauri has a speed that is more noticeable than c due to the fact that the velocity at the edges of the object that is moving is due to the distance and the rakish speed. Additionally, it's possible that objects, like comets to change their speed from superluminal to subluminal, and vice versa in part due to the fact that their distance from Earth is variable. Comets could have circular orbits that extend out up to 1000 AU. The circumference of a circle that has the range that is 1000AU may be much more impressive than a single lighter day. In the final morning, the comet located at that distance can be superluminal with a geostatic and thus it is non-inertial and has a non-linear outline.

Shadows and light spots:

In the event that laser beams are cleared of a removed object the laser light may without much effort be directed to travel across the item at a rate that is greater than c.[7]In the same way the shadow projected onto an unaccessible object could be set to travel over the object quicker than c.[7]In neither of these cases does the light travel from the source to the object faster than c, nor do any other data move faster than light. [7][8][9]

There can be any "impediment" (or deviation) of the visible location of the wellspring the electric or gravitational field when the source is moving at a consistent rate and frequency, this static field "impact" could appear at first glance as if it has been "transmitted" faster than light's speed. However there is a uniformity of movement that is generated by the source could be eliminated by adjusting in the reference outline, which brings the field's bearing static field to shift quickly regardless of the distance. It is not a shift of position that "engenders" an effect, so this kind of change isn't used to transfer information directly from source. Any data or object cannot be transmitted or engendered via FTL from the source to the recipient/eyewitness via the electromagnetic fields.

Speeds of shutting:

The speed of two frames of reference moving within a single casing of reference become closer to each other is referred to as the shutting or shared speed. This could exceed the speed of light because of two frames of reference moving in the same direction as light in inverse bearings , as for the outline of the reference.

Imagine two pframes that are swiftly moving references drawing closer from opposite sides of an collider-like atom smasher type. The velocity at the end would be the speed at which separation between the two references diminishes. From the point of view of an eyewitness who is close to the quickening agent the rate is not even close to two times the speed of light.

Relativity does not prohibit this. It informs us that it's not appropriate to use Galilean relativity to measure that speed for one pframes that would be observed by an eyewitness being close to another molecules. In other words, extraordinary relativity can provide the appropriate equation to calculate such a speed.

It's instructive to determine the speed and relative velocity of pframes that move at v and - V in the quickening agent outline and compare it to the velocity at which they end $2v > C$. Communication of the speed in terms of c units, the equation $ss = v/c$

If spacecrafts travel to a planet that is one light-year (as determined by the rest of Earth's outline) away from Earth at a rapid speed pace, the time it takes to reach that planet may be less than what is one year, as measured by an timer of the explorer (in spite of it will always be more than one year according to the timer on Earth). The efficiency achieved by partitioning the space through the Earth's case, when measured by the clock of the voyager is called an appropriate rate or legitimate speed. There is no limitation in the calculation of an appropriate rate since legitimate speeds do not refer to a speed that is measured by a single inertial edge. A light flag which left the Earth while the traveler was bound to reach the destination earlier than the traveler.

As one cannot move faster than light speed it is possible to conclude that humans will never be further away from Earth than 40 light years if the person who is traveling is between the ages of 20 to 60. A explorer will never be able to reach more than the numerous star structures that exist within the range of 20-40 light years from Earth. This is a contradictory conclusion: because of increasing time that the explorer is able to travel many light-years over their 40 dynamic years. If the spacecraft speeds up at an average of 1.25 millimeters (in its own shifting edges of reference) then it will, in the course of 354 days, attain speeds close to the speed of light (for an observer at Earth) and also time enlargement can extend their life span to a massive amount of Earth years, as seen from the perspective of The Solar System, however the duration of the voyager's personal life won't follow the same lines. If the traveler returns to Earth and returns, they will be at a significant time in the future of Earth. Their speed won't be seen as being higher than the speed of light by the people who watch on Earth as well as the explorer will not quantify their speed as being greater than the speed of light however, they will notice a constriction in the length of the universe towards them of traveling. Furthermore, when the explorer turns to give back the Earth appears to be experiencing an extended period of time than the traveler is experiencing. In spite of being aware that the (normal) speed cannot exceed c but the four-speed (separation that is seen by Earth isolating them by their own, i.e. subjective time) is often more noticeable than the speed of c. This is evident in the measurable studies of muons that travel much further than the half-life of c (very still) or flying close to c.[1010

Stage speeds that are higher than the speed of c

The speed of a stage electromagnetic wave, as it travels through a medium often exceed c, the speed of light in vacuum. This is the case in a variety of glasses at the X-beam frequency. But the speed of the stage of a wave can be correlated to the rate of spread of a possible singular-recurrence (absolutely unichromatic) section of the wave in the frequency of that time. This segment of the wave must be continuous in its degree and have a consistent degree of quality (else it's not monochromatic) which means it is unable to transmit any information. Therefore, an increase in speed over c doesn't indicate the spreading of signs at the speed of c.[1313.

Bunch speeds are higher than the speed of

The speed at which an oscillation (e.g. an elongated shaft) can also surpass c under certain conditions. In these instances where the conditions involve a rapid diminution of force the largest portion of the envelopes of a heartbeat might be at a higher speed than that of. In any event it is not a sign of the spread of signs that have speeds that exceed c, regardless of the fact that you might be enticed by the idea of join beat maxima by signals. This last association has seen to be deceiving, principally because of the fact that data upon the arrival of a heartbeat may be gathered before the beat that is the most powerful arrives. For instance, in the possibility that a component allows the full transmission of primary portion of a heartbeat, while massively weakening the beat that is most severe and all that follows (contortion) in which case the beat with the highest frequency is transferred forward, and the information on the beat is not faster than c, despite this impact. But the speed of assembly can be higher than the speed of c within a couple of parts of the Gaussian shaft that is in the vacuum (without diminution). The diffraction results in the heartbeat's top frequency increases more quickly, while general force doesn't.

All-inclusive expansion

The universe's history - gravitational waves may arise from massive expansion, and a faster than light extension that occurs shortly in the aftermath of the Big Bang (17 March 2014). [18][19][20]

The expanding universe can cause distant systems to move away from us faster than light, when proper separation and cosmological time are used to calculate the speed of these universes. Whatever the case it is true that, in the context of relativity is a local concept, therefore speed calculated using comoving arrangements doesn't have any fundamental relationship to the speed that can be measured locally. [21The concept of speed is not a local one. (See comoving separation to see the exchange of ideas regarding "speed" in the field of cosmology.) Rules applicable to speed ratios in singular relativity, like the notion that relative speeds shouldn't increase beyond the speed of light, do not have any significance to the relative speed of comoving facilites which are typically depicted in terms of being the "development in space" across universes. This rate of expansion is believed to have reached its peak during the time of inflation, which is believed to have occurred in the second period following the Big Bang (models propose the timeframe would be from 10 to 36 seconds after the Big Bang to around 10-33 seconds) in which case the universe could have expanded by an amount of about 1020-1030. [22]

There are a variety of systems visible in telescopes with red motion amounts of 1.4 or more. They are currently moving away from us at speeds that are greater than the speed of light. Because that the Hubble parameter is shrinking over time, there could exist situations in which the world is moving away faster than light is able to emit an indication that reaches us at some point. [23][24]

"Our most compelling molecule skyline" is the massive microwave foundation (CMB) at redshift at z > 1100 due to the fact that we cannot see beyond the surface of the last diffusing. Even though the surface of last diffusing isn't at any alteration of comoving coordinate, the current slowdown of the focuses from which CMB has been transmitted was 3.2c. In the time of outflow, the rate was 58.1c which was accepted as (OM,O?) = (0.3,0.7). We regularly watch protests that retreat faster than the speed of light as well as it is clear that the Hubble circularity isn't an equilateral triangle." [25]

However, given that the evolution in the Universe is speeding up the speed of development, it is predicted that all systems will at the final phase cross a type of cosmological event skyline, where any light that is released beyond that point will not be able to communicate with us in the foreseeable future, due to the fact that light does not reach a level that it's "impossible to be missed speed" towards us is greater than the distance speed that extends away from us (these two ideas of speed are also thought about in Comoving distance#Uses to determine the right separation). The distance between us and the cosmological event skyline is about 16 billion light years, which implies that a signal of an event that is taking place today could eventually be able to communicate with us in the future when the event was only 16 billion light-years from us but it would never be in touch with us if that event was further than 16 billion light years away. [24]

Galactic observations

The apparent superluminal motion is observed in many radio universes and blazars quarks, and as well in microquasars. The phenomenon was anticipated long prior to being observed by Martin Rees[clarification requiredand is explained as an optical illusion caused by the object slightly moving towards the observer,[27even though the velocity numbers as a fact. The phenomenon does not invalidate the theory of relativity. Strangely enough, the revised figures show that these frame references move at close to the speed of light (with regard to the outline of our reference). This is the main instance of a large mass moving at a speed of light. The Earth-bound research centres are now able to speed up small amounts of pframe reference references that are basic to these speeds.

Quantum mechanics

Certain quantum mechanics marvels such as quantum snare can provide the illusion that they allow data correspondence more quickly than light. Based on the hypothesis of no-correspondence, these marvels do not permit real correspondence, they let two people in different areas view the same framework at the same time, and there is no means of controlling what they see. Wavefunction breakdown is seen as an example of quantum decoherence. It is simply an impact of the invisible neighborhood time progress of the wavefunction in an object and the greater part of its environment. Since the basic principle doesn't negate causality within the neighborhood or permit FTL, it follows that as does the additional effect of wavefunction breakdown whether it is genuine or not.

The vulnerability standard implies that photons could travel for small distances with velocities that are certain degrees faster (or more slowly) than c in a vacuum. This possibility must be considered when trying to identify Feynman charts of an interaction between molecules. It was revealed at the end of 2011 that single photon might not travel faster than c.[30 In quantum mechanics, the virtual reference to pframes may travel faster than light. This phenomenon is explained by the idea static field impacts (which are mitigated by virtual reference to pframes using quantum terminology) might travel faster than light (see the section for static field impacts above). In any event, it is evident that these vacillations are common and are aimed at proving that photons actually travel straight lines through long (i.e. non-quantum) distances. Furthermore, they are moving at the speed of light, by and large. Therefore, this doesn't indicate the possibility of superluminal data transmission.

There have been a variety of reports in the renowned press about faster than-light transmission in opticsoften in relation to the quantum burrowing marvel. In general, these reports have the stage speed or gather speed that is faster than the speed of light in a vacuum. But, as mentioned above, a superluminal stage speed is not a suitable method to speed up the transmission of information. There's once moment been a tense debate over the final point. A channel that permits such spread isn't able to be created more quickly than the speed of light.

Quantum teleportation sends quantum information in any speed that is used for transmission of the exact quantity of conventional data, possibly the speed of light. Quantum data could theoretically be utilized in one of ways traditional data is not be utilized, for instance, using quantum calculations that include quantum data that is only accessible to the user.

Hartman effect

The Hartman impact is a burrowing effect through an area where the burrowing speed increases to a constant rate for massive barriers. This was first drawn in the work of Thomas Hartman in 1962. [32] It could be the crevice that exists between two crystals. At the moment that crystals are in contact, light passes straight through. However, if there's a hole the light is reflected. There is a high probability that the photon will travel into the crevice instead of to follow the refracted route. If there are large crevices between crystals, the time of burrowing is an even pace and the photons appear to have crossed at an extremely high speed. [33]

In any event the study conducted of Herbert G. Winful from the University of Michigan recommends that the Hartman impact cannot be used to harm the relativity of signals by transmitting them faster than c, because of duration of the burrow "ought not be tied to the speed at which flashing waves that don't travel". The transitory waves that occur in the Hartman impact are caused by virtual pframe reference and an unengendering static field as defined in the sections below for electromagnetism and gravity.

Casimir effect

In the field of science in science, the Casimir impact, also known as the Casimir-Polder power is a physical energy that is transferred between two objects due to the reverberation of energy from vacuum within the interceding space frames. It is often depicted in the form of virtual pframes that cooperate with frame references, which is inferred from the scientific form of one possible method to measure the impact's quality. the impact. Because the power's power quality decreases rapidly when separated It is only quantifiable when the gap between the frames is a large extent small. Since the cause is of virtual pframe reference interfering the static field and is susceptible to the comments regarding static fields discussed in the previous paragraph.

EPR paradox

The EPR mystery is a reference to a famous investigation of Einstein, Podolski and Rosen which was formally acknowledged through Alain Aspect in 1981 and 1982 during his Aspect test. In this test the assessment of the conditions for one quantum framework in the caught combine clearly restricts the other one (which could be quite different) which is measured at the appropriate state. However, there is no way for data to be sent along these lines The response regardless of whether the estimate has any effect on the quantum framework in question depends on the quantum mechanics interpretation one is a part of.

An investigation conducted during 1997 by Nicolas Gisin at the University of Geneva has shown non-nearby quantum relations between Pframe references that are separated from each other by between 10 and 20 km. As mentioned earlier, the non-neighborhood links discovered in entrapment cannot be used to transfer established data faster than light, and so relativistic causality is not destroyed as a no-correspondence theory. more information. A quantum science study in 2008 also conducted by Nicolas Gisin and his partners in Geneva, Switzerland has confirmed that in any speculation about non-nearby-hidden variables hypothesis, the speed of quantum non-neighborhood connection (what Einstein called "spooky activity at a distance") is not less than 10,000 times the speed of light. [36]

Postponed decision quantum eraser

Postponed quantum eraser (a test by Marlan Scully) is a reinterpretation of an

EPR Catch 22 in which the perception or absence of impedance following the entrance of a photon by a twofold test of opening depends on the state that a secondary photon that is trapped by the primary. The standard for this test is that perception by the 2nd image may occur later that the initial perception initial photon. This could give an impression that the estimate of the subsequent photons "retroactively" determines whether prior photons reveal obstruction or not however the impedance scenario can only be observed by connecting the estimates of both people in the pair, and therefore it cannot be visible until both photons been evaluated, which guarantees that the person who is observing only those photons that experience the opening will not receive information on alternate photons when they are when using a FTL or reverse in-time. [38][39]

Superluminal communication

It was suggested that Superluminal correspondence should be merged in this report. (Talk about) Since April 15, 2015.

The speedier than-light correspondence is, as per Einstein's theory of relativity, proportional to the speed of travel. Based on Einstein's idea of unique relativity, the quantity we consider to be the speed of light in the absence of a vacuum (or close to a vacuum) is in fact the primary physical consistency c. This means that every inertial observer, paying no thought to the speed they are traveling at and mass, will always measure the pframe reference of zero mass like photons traveling at c speed in the vacuum. This means that estimates of speed and time within different casings aren't more connected than continuous movements, but are actually influenced through Poincare shifts. These changes can have significant implications:

The force of a relativistic gigantic molecule would increase by pace, in a manner that at the speed of light, an object could have unlimited energy.

To accelerate an object with non-zero rest mass will require endless time, with some speeding up or an exponential increase in speed over a certain amount of time.

In any case, acceleration requires endless energy.

Some eyewitnesses who have sub-light relative movements will disagree on which occurs first in two instances that are separated by a space-like period. In the final analysis all day long, any journey that is more rapid than light can be interpreted as traveling backwards in time within another, equally significant or reference casing, or the need to be able to accept the theory of possible Lorentz violations without further discussion on a secret scale (for instance, this is the Planck size). So, any theory that provides "genuine" FTL additionally needs to be able to accommodate to time travel and related paradoxes, or at least to assume that the Lorentz Invariance as a symmetry in thermodynamical quantifiable nature (henceforth the symmetry is broken by a moment of, but not noticeable on a scale).

In relativity, the direction speed of light is only guaranteed to be c when it is an inertial edge. However, in an inertial edge, the speed of light in the direction may differ as c.[43All in the realm of relativity, no direction framework for a large area of spacetime bended is "inertial" therefore it is possible to use an international direction framework in which objects travel faster than but in the immediate area of any point within spacetime that is bended, we can define as a "nearby inertial line" and the local speed of light is at c in this frame with huge objects traversing the neighborhood at an average speed of c that is not as high as the rate of light in the local inertial casing.

Justifications

Einstein's relativity conditions suggest that the speed of light in the (close) space is constant for cases of inertial force. In other words, it will be the same for any reference casing moving at a steady speed. The parameters don't reveal any specific quality of the speed of light, but it is a tentatively determined amount for an unresolved length unit. Since 1983 the SI length unit (the meter) has been defined using the speed of light.

The trial verdict was conducted in a vacuum. However that the vacuum we have is not the only vacuum that exists. The vacuum is a source of energy to it, known as the vacuum energy, and it may be altered in certain circumstances. When the energy from the vacuum is decreased the light has been predicted to move faster than the normal value of in c. This is also known as"the Scharnhorst impact. This type of vacuum is made by bringing two perfectly smooth metal plates together with close distance to the nucleus' breadth of division. It's referred to as the Casimir vacuum. The results of Counts indicate that light will accelerate in such the vacuum with an infinitum sum. A photon traveling between two plates one micrometer apart would increase the speed of the photon by 1 section in 1036. In other words it has until the present time no evidence to support an exploration of the predictions. A more recent analysis[47] suggested that Scharnhorst's impact Scharnhorst impact cannot be used to transmit data backwards in time. A single plate arrangement that is later than the rest casing of the plates, which could be as a "favored edge" to use for FTL flagging. However, when there were multiple sets of plates moving with respect to each other, the authors discovered that there were no arguments which could "promise that the total absence of causality violation" and referred to Hawking's theoretical order security assumption that suggests that the criticism circles of virtual pframe references could result in "wild peculiarities in the quantum stress-energy renormalized" in the case of any time machine that could be built that will require a hypothesis of quantum gravity that is difficult to investigate. There are various theories that claim that Scharnhorst's original study, which appeared to show the possibility of signals that were faster than c, contained estimates that could be inaccurate in the hope that it's not entirely clear whether the effect could actually increase signal speed in any way. [48]

However, an investigation conducted of Herbert G. Winful from the University of Michigan recommends that the Hartman impact cannot be used to alter the relativity of the universe by transmitting signals faster than c on the ground that the burrowing period "ought not be correlated with the speed at which flashing waves that don't spread". The transient waveforms that are observed in the Hartman impact result from virtual pframe references as well as an unproliferating static field as described in the sections previously mentioned for electromagnetism and gravity.

Casimir effect

In the realm of science in science, the Casimir impact, also called Casimir-Polder power can be described as physical power that is applied to certain objects because of the reflection of vacuum energy within the space that mediates between frame references. This is one of the instances depicted with regard to virtual pframes that interface with frame references that can be inferred from the scientific nature of one method that is conceivable for determining the caliber of impact. Because the power's quality is reduced rapidly with separation the frame references, it can only be quantified when the gap between the frames is very little. Since the cause is of virtual pframe reference interfering an impact of a static field it could be a result of the comments about static fields discussed in the past.

EPR paradox

The EPR mystery refers to a well-known study by Einstein, Podolski and Rosen which was formally acknowledged through Alain Aspect in 1981 and 1982 during his Aspect test. In this study the assessment of the state that one framework of quantum mechanics that comprise an entrapped combination clearly limits the other framework (which may be unaccessible) to be assessed in the reciprocal state. However, there is no way that information can be sent in this manner, and the decision whether the estimate actually affects the other quantum framework depends on which knowledge the quantum theory one adheres to.

A study conducted during 1997 by Nicolas Gisin at the University of Geneva has revealed non-nearby quantum connections between Pframe references that are separated from each other by 10 km. However, as we've mentioned previously, the non-neighborhood connections discovered in trap cannot be used to transfer the data in the same speed as light, which is why causality in relativity is not affected by the no-correspondence theory. See no-correspondence for more data. A quantum science experiment conducted in 2008 was also carried out by Nicolas Gisin and his partners in Geneva, Switzerland has verified that in any non-neighborhood hypothesis, even speculative shrouded variables theory, the speed of the quantum non-nearby connection (what Einstein called "spooky activity at a distance") is not less than 10,000 times the speed of light. [36]

Deferred decision quantum eraser

A delayed decision-quantum eraser (an investigation of Marlan Scully) is a variant that is a variant of EPR Catch 22 in which the perception of or lack of obstruction following the division of a photon in an opening test that is twofold depends on the state of perception of the second photon that is caught by the first. The norm procedure for this type of test is to have the perception of the secondary photon could occur later as the first initial photon. This could give the impression that the estimate of the subsequent photons "retroactively" determines if prior photons have resistance or not, despite being aware that an example of obstruction has to be observed by comparing the assessments of both participants from the pair, and thus cannot be observed until both photons are assessed, ensuring that an observer who is only observing the photons that are experiencing the opening doesn't get information about the other photons either whether in a FTL, or reverse in-time. [38][39]

Superluminal communication

It was suggested that Superluminal correspondence should be merged in this report. (Talk about) Since April 15, 2015.

The faster than-light correspondence is, as per Einstein's hypotheses of relativity, similar to time travel. Based on Einstein's theories of relativity, the measurement we use as the speed of light in the absence of a vacuum (or close to it) is actually the fundamental physical consistency c. This means that every inertial eyewitness paying no attention to their speed will always quantify zero-mass references to pframes, for instance photons moving at C in an atmosphere. This implies that estimates of speed and time in different edges are not connected only by constant shifts, but are instead connected by changes in the Poincare equation. These changes have profound implications:

The force of a relativistic massive molecule would increase as velocity increases in a manner that at the speed of light an object could have unlimited energy.

To accelerate an object with non-zero rest mass requires unbounded time at the possibility of a limit on speed or never-ending speeding up for the duration of a time limit.

In any case, the rapid growth requires constant energy.

Some observers with sub-light-like relative movements are not sure which occurs the first time in any two instances that are separated by the space-like interval. In other words the case, any movement that is more rapid than light can be interpreted as traveling through time in reverse another, equally legitimate casings of reference or the possibility of Lorentz violation within a few seconds at a the scale of a secret (for instance, for instance, the Planck scale). Thus, any theory that permits "genuine" FTL likewise needs to be able to accommodate to time travel and its associated paradoxes, or at least acknowledge the Lorentz invariance as an invariant symmetry of thermodynamical nature (henceforth the symmetry is broken in a short time at a scale that is not easily discernible).

In the unique case of relativity, the speed of light's direction is only guaranteed to be c within an inertial edge. However, in an inertial casing, the speed of light's direction could be different from c.[43When all is said and done relativity, no direction framework for a large region of bended spacetime is "inertial" and therefore it is possible to use a global direction framework where objects move faster than c but in the immediate vicinity of any point in spacetime that is bended, we can define the term "nearby inertial casing" and the local speed of light would be at c, with massive objects moving through the neighborhood continuously having the same speed as c within the local inertial casing.

Justifications

Einstein's relativity conditions suggest that the speed of light in the (close) space is constant for the inertial edges. This means that it will be the same for any other reference frame that is moving at a constant speed. The circumstances don't give an exact value for the speed of light, but it is a tentatively determined value for a modified length unit. Since 1983 the SI length unit (the meters) is characterized using the speed of light.

The test's determination was done in vacuum. However it is clear that the vacuum we are familiar with isn't by any means the only vacuum that exists. It has energy associated to it, referred to as the energy of the vacuum that could be altered in certain situations. If the energy of the vacuum is reduced it is expected to accelerate faster than standard high-quality C. This is called"the Scharnhorst impact. This type of vacuum is created by bringing two extremely smooth metal plates together with close distances to divide the nuclear breadth. It's referred to as the Casimir vacuum. The theory is that light will accelerate in such the vacuum due to the microscopic amount: A photon traveling between two plates that are one millimeter apart would increase the speed of the photon by 1 section in 1036. So it is believed that there has until this point been no testing of the prediction. An analysis from a few years ago suggested that Scharnhorst impact cannot be used to transmit data backwards to time. A single plate arrangement that is later than the edges of the plates. This could be an "favored case" to allow FTL flagging. With various sets of plates in motion relative to one another, the authors noted that there were no arguments which could "promise the absence of causality violation" and referred to Hawking's theoretical insurance guess which suggests that circles of criticism of virtual pframe references could create "wild peculiarities in the renormalized quantum stress energy" in the case of any time machine that could be built in this way, and will require a theory that quantum gravity is a reality to dissolve. There are various theories that claim that Scharnhorst's analysis, which seemed to show the possibility of higher-speed signals included estimates that might be wrong in the hope that it's unclear what the effect of this could be to create signal speed in any way. [48]

In some theories with breaking Lorentz symmetry, it's speculated that the symmetry retained in the fundamental scientific principles, but the unconstrained breaking of symmetry Lorentz invariance [66] shortly following it was triggered by the Big Bang could have left an "relic field" across the universe that causes references to pframes to continue in a variety of ways, based on their speed relative to the field.however, be that way it is also possible to find certain models in which Lorentz symmetry is softened in through a different method. If Lorentz symmetry is not able as a fundamental theorem at Planck scale, or at other critical scale it is possible that pframe references that have an initial velocity that is not identical to the speed of light are the definitive element of matter.

In the present theories of Lorentz symmetry violation the phenomenological parameters have been considered being energy dependent. Therefore, it is generally accepted that [68][69the existing limits for low energy cannot be attributed to wonders of high energy and, in any event there are numerous investigations into Lorentz violation at high energies have been conducted using methods like the Standard-Model Extension. [65The reason for this is that Lorentz symmetry violation is needed to be grounded when one moves closer towards the center scale.

Another theory that has been proposed in the past (see EPR Catch 22 above) resulted from the investigation of the EPR correspondence with the fundamental device for removing the extremely impeded terms of the Lorentz change to give the most popular superior reference frame. The casing cannot be used to conduct research (i.e. analyze the effect of light-speed limited signals) but it provides an end-to-end, straight edge that all can agree upon If superluminal correspondence was possible. In the event that this seems liberal and allows for cooperation with all space and time and an determined universe (alongside the decoherence hypothesis) in contrast to business as usual, which permits time travel and causality conundrums as well as subjectivity during the estimation process, and different universes.

Hypotheses about physical vacuum that are superfluid

In this view, the vacuum in which we live is viewed as a quantum superfluid that is essentially non-relativistic, whereas the Lorentz symmetry isn't an exact symmetry of nature but rather the imperfect representation of small changes in the superfluid's background. In the structure of the theory, a hypotheses was suggested in which the physical vacuum was interpreted to represent that of the quantum Bose fluid, whose ground-state function is represented by the Schrodinger logarithm. It was discovered by the results that the relativistic gravity connection is a result of the little adequacy aggregate excitation model[73] whereas relativistic rudimentary pframe references can be represented by the molecule-like modes at the most extreme of low-momenta. The crucial fact is that at very high speeds, the behavior of molecules like modes can be distinct from the relativistic one. they can reach the speed of light's furthest point, but with only a small amount of energy. Moreover it is possible to generate more than light faster than possible without the need for reference to moving frames that be of an insignificant mass. [75][76]

The time of flight of neutrinos

MINOS experiment

In 2007, the MINOS collaboration published results measuring the flight time for 3 GeV neutrinos, which produced a speed which was faster than light with 1.8-sigma significance. But, these estimates were thought to be measurably stable and neutrinos were moving in the same direction as light. [78] Following the fact that the discoverers of the project were revised at the end of 2012 MINOS corrected their original results and found a correlation with light speed. Additional estimations will be made. [79]

Musical show, neutrino anomaly

On the 22nd of September in 2011, a report [80published by the OPERA Collaboration discovered the 17th and 28th GeV muon neutrinos that traveled 730 km (454 miles) from CERN close to Geneva, Switzerland to the Gran Sasso National Laboratory in Italy traveling faster than light, based on a ratio measurement that was 2.48×10^{-5} (around one in 40000) an amount that has 6.0-sigma significance. On the 18th of November 2011 a follow-up catch up test conducted by OPERA researchers confirmed their original findings. But, the researchers were skeptical of the effects from these experiments, and the validity of which was debated. In March 2012 in March 2012, the ICARUS group failed to replicate the OPERA results using their equipment and was unable to distinguish neutrino venture duration across CERN and CERN to Gran Sasso National Laboratory indistinct from the speed of light. In the following months, the OPERA group revealed two flaws in their setup for their hardware which led to blunders which were way beyond their unique in the interim: a fiber optic link that was not connected properly, which caused the clearly higher than light estimations, as well as an oscillator clocking too quickly. [86]

Tachyons

In the case of singular relativity, it's difficult to accelerate an object to the speed of lightor for an huge objects to move at the speed of light. In any case it is possible for an object to exist that always moves faster than light. The pframe reference theories with this feature are referred to as Tachyonic pframe reference. They attempt to quantify them to make faster than light reference pframes, and instead explained that their proximity triggers an instability. [87][88]

Different scholars have recommended that the neutrino may have a tachyonic nature,[89][90][91][92][93] while others have debated the possibility. [94]

General relativity was developed after relativity in order to include ideas such as gravity. It abides by the principle that any object cannot speed up to speed equal to light at the edge of an incidental observer. But it permits variations in spacetime that allow objects to move faster than light when viewed from the perspective of an unobserved observer. One of these mutilations is the Alcubierre drive. This could be viewed as delivering the spacetime to expand and transports an object along. Another possible framework is the wormhole that connects two distant zones in the same way as an alternative route. Both bends need to form an extremely strong arch within a very narrow space-time area the gravity field of both would also be enormous. To mitigate the vulnerability and prevent the bends from collapsing in their own 'weight' the user would have to create theories of exotic matter, (or negative energy).

General relativity further suggests that any technique for faster than light travel could also be employed to speed up time travel. This raises concerns about causality. A lot of physicists believe these wonders are not possible and that future speculations about gravity will deny the possibility of. One theory states that steady wormholes could be possible but any attempt to make use of the wormhole system to ignore causality could result in their destruction. In string hypothesis, Eric G. Gimon and Petr Horava have argued[97] that in a supersymmetric five-dimensional Godel universe, quantum redresses to general relativity viably cut off districts of spacetime with causality-disregarding shut timelike bends. Particularly in the quantum hypothesis, there is a spread supertube which cuts spacetime in such a way that, despite it being true that throughout complete spacetime a time-like bend passed through each point, no completely bent bends can be found in the area that is limited to the tubes.

Light intensity variable

In the realm of science, the rate of light within a vacuum believed to be consistent. There are theories that support the speed of light being a constant. The explanation for this announcement can be described as follows: the following.

The speed that light travels through is a dimensional quantity and, as was pointed out in this context by Joao Magueijo. It cannot be measured. Quantities that can be measured in the field of material science are, in all respects they are, in spite being that they're often created as proportions of dimension quantities. For example, when the height of the mountain gets determined in the real world, what is being determined is the ratio of its height with the measurement of the meters stick. The conventional SI system of units is based upon seven primary dimensional values that include separation of time, mass electrical present, thermodynamic temperature, measurement of substance, and radiative intensity. The units are considered to be free and therefore they aren't able to be represented with respect to each other. Another option is making use of a specific set of unit arrangements, it is possible to reduce the estimations to dimensional numbers that communicate proportions between the quantities being measured and various central constants, such as Newton's steady speed of light, and Planck's constant physics can define as little as 26 dimensionless constants that can be communicated in such proportions and are considered to be independent of each other. In managing the basic dimensional constants, one can also develop Planck length, time and Planck energy. Planck duration and Planck energy that form a suitable arrangement of units to communicate dimensions, also called Planck units.

Magueijo's idea utilized an alternative configuration of the units. This is a move which he defends by stating the argument that certain circumstances will be simpler with these new units. With the new unit, he sets with the fine structure constant, a quantity that some individuals, using units as a component of where the speed of light changes and argued is time-subordinate. In this manner, when units are arranged where the fine structure consistency changes, the observed scenario is that the speed of light is subordinate to time.

Chapter 8: Time Dilation

Time dilation (here called "time wideding"/"time expansion") is the reason why two functioning watches report distinct times following various increases in speeds. For example, ISS space explorers come returning from their missions just a bit less than they would have had if they were on Earth as well. GPS satellites function because they are compatible with bowing of spacetime, which allows them to work with the frameworks of Earth. [1]

In the relativity hypothesis the concept of time enlargement can be described as an enlargement of the time between two events that are observed by observers either moving in relation to one others or contrastily arranged in relation to a gravitational mass or masses.

A clock that is very quiet for one observer could be found to tick at a different speed when compared with the clock of a second person watching. The reason for this is not due to special parts of tickers or from the increasing times of the signs, but due to the nature of spacetime.

The timers that are on board Space Shuttle Space Shuttle run marginally slower than the timekeepers that are used in reference on Earth and the clocks of GPS or Galileo satellites are a bit quicker. This time-widening is repeatedly shown (see the test affirmation beneath) as a result of small abberations of nuclear checks on Earth and in space regardless of the fact that both timekeepers function flawlessly (it isn't an error in the mechanical system). The nature of spacetime is such that the measurement of time across a variety of directions is affected by the contrasts between gravity or speed , each of which affects time in a variety of ways. [2][3]

In theory in order to provide an easier illustration the time expansion may affect scheduled gatherings of space travelers featuring cutting-edge technologies and even more impressive speed of travel. Space explorers will need to establish their timekeepers to track at least 80 years. Meanwhile, the control of missions - back on Earth might have to count the years to 81. Space travelers will come back to Earth following their primary purpose, and have matured by in one year but not quite the average population, but still focus. In addition, the experience of time passing is a change for anyone. So space explorers on the vessel and the mission control staff on Earth are all the same despite the impact of the speed of time increasing (i.e. for the people who are voyaging those who are stationary live "quicker" and for those who have stopped, their counterparts who are moving live "slower" in every minute).

Innovations are limiting the speed of space travellers and space travelers, the distinctions between them are nil in the following six months of the International Space Station (ISS) The space explorers group is not like the ones on Earth but only by 0.005 seconds (no comparable to the one-year difference from the hypothetical scenario). The impact are more significant when space travelers were traveling at the speed of light (299,792,458 meters per second) and not their true speed - that is, the speed of the speed of the ISS 7700 meters per second. [4]

The increase in time is caused through contrasts in gravity or speed. Due to ISS it is slow due to the speed of the roundabout circles; this effect is reduced by limitation of the gravitational potential.

Time dilation at a relative speed:

When two people are in motion that is unaffected by any gravitational force and the view of both is in the sense that their (moving) clock running at a lower rate than the clock in the nearby. The more rapid the speed increases, the more notable the degree of time expansion. This phenomenon is often known as exceptional relativistic time expansion.

For instance two rocket transports (An as well as B) moving at a high speed in space could experience the issue of time expansion. If they in one way or the other had a fair view of each other's vessels and could be able to see each other's development and timekeepers as moving much more slowly. This means that, within the edges of the horizon of Ship An all is going normally, but everything the top of

Ship B has all the indications of going more slowly (and in reverse).

From a distant perspective from a distance, time-based enumeration by timekeepers which are still in relation to the edge of the neighborhood of the reference (and quite a distance of any gravity mass) always seems to move at the same speed. In other words, if another ship like Ship C moves in close proximity to Ship An, it is "very still" in relation the Ship A. From the viewpoint of Ship A, another Ship C's chances would appear to be normal, too. [5]

An issue arises when ship An A and Ship B are both of the opinion that each that the other's chances are less or more slowly, which one will have gotten significantly in the event they choose to be together? With a greater understanding of time-widening relative speeds this appears to be a the twin Catch 22 turns out not to be an anomaly in any way (the solution to the mystery is based on a jump in time, which is an effect of the speedy moving of the spectator). Therefore, understanding that the two Catch 22 would clarify why astronauts who are on the ISS are getting older (e.g. 0.007 seconds behind frequent intervals) even though they are experiencing the enlargement of relative speed.

Gravitational time dilation

Time moves rapidly away from the focal point of gravity as observed by huge frames (like Earth). Earth)

The main reason is that both witnesses are distinctly placed with respect to their distance from a massive gravitational mass. The general theory of relativity describes the fact that for both observers the clock closest to the gravitational field, i.e. more deep with respect to it's "gravity well" appears to move slower than the clock which is further away from it. It is a result of the opposite effect of the speed relative and its enlargement.

Gravitational time expansion has an impact on everything e.g. for ISS space travellers. For ground observers, the ISS space travellers are at a relative speed that slows their speed, while the reduced gravitational force at their locations accelerates it. The two impacts that limit their impact aren't equally solid. At ISS top, the net impact is back-off of timekeepers even though in circles with higher elevations, tickers move faster than they do on the ground.

This effect is not restricted to space travelers and a climber's chance is moving a little faster at the highest point an elevation (a high elevation, and further away from earth's center of gravity) as compared to people who travel at the in the adrift zone. It's also been calculated that due to the fact that time is expanding the middle of Earth has a lifespan of 2.5 years older that the surface. [6]

Similar to untouched expansion, the sensation of time is common (no any one notice of a distinct in their own particular frame that they refer to). In the case of increasing speed, the two observers observed the clocks as moving at a slower rate (a similar impact). Now, with gravitational time expanding, both observers who are at the bottom, as opposed to the climber - agree that the clock nearer the mass is slow in speed, and both accept the degree of the difference (time growth due to gravity is not in a way complementary). The climber perceives the ocean-level timekeepers moving more slowly, while those who live at adrift level view that the clock in the climber's is speedier.

In Einstein's theory of relativity, the time enlargement in these two situations can be reduced to:

In relativity (or it is speculatively a lengthy distance from all gravitational mass) the tickers moving in relation to an inertial configuration of perception are found to be running slowly. The impact is accurately depicted by the Lorentz shift.

In general relativity, time in a location with a smaller gravitational energy - for instance, those in close proximity to a planet are observed to be running at a slower pace. Frame references on gravityal time enlargement as well as redshift of gravitation provide a more specific exchange.

General relativistic and extraordinary impacts can be joined (as observed by ISS space travellers).

In relativity the time enlargement effect is equal. If you view it from the viewpoint of both checks which are in motion with respect to each one, it is the second clock that has been time-enlarged. (This assumes that the relative motion between the two sides are the same which means that they do not speed up relative to each in the course of their perceptions.) It is interesting to note that gravitational time expansion (as is viewed in the context of all as a matter of relativity) doesn't have to be proportional. those who are onlookers at the top level of a tower would be able to see timekeepers in the ground are ticking slower, while those who are at ground level will agree regarding the direction of the clock and the magnitude of the difference. However, there is a slight differences because all of the observers believes that their local timekeepers However, the direction and proportion of gravitational increase is shared by all observers, regardless of their height.

Sci-fi implications:

Science fiction fans have been noticing the effects that time expansion has a positive impact on advance time travel and in actual making it feasible. 7] The Hafele test and Keating test involved flying planes across the globe equipped with nuclear timekeepers that were loaded up. When the treks' climax was reached, the tickers were displayed in contrast to the static, ground-based nuclear clock. It was discovered that 273+7 nanoseconds were recorded in the plane's clocks. The current record-holder for the human time travel record is Russian cosmonaut Sergei K. Krikalev, who beat the previous record of 20 milliseconds held by Cosmonaut Sergei Avdeyev. [10]

The basic induction of time expansion due to the relative velocity

Eyewitness is still able to measure time $2L/c$.

Eyewitness that is moving in a parallel manner to the setup is able to measure longer distance and consequently, at the same speed of light c, is able to measure longer time. $2D/c$ is greater than $2L/c$.

An: the area of the base mirror, when sign is generated at time when t'=0.

B: the area on top of the mirror where sign is reflected when D/C =.

C: the area of the reflection of the base when sign appears in a state of crisis at the time 2D/c.

Time-widening can be derived from the observed stability of light's speed in every reference frame. [11][12][13][14]

The consistency in the speed of light indicates, contrary to our instincts, that the paces of light and material questions are not a source of added to the substance. It is not practical to make the speed of light appear more notable by advancing closer toward the source of the material which is transmitting light. It is not practical to make the speed of light appear smaller by moving away to the point of origin at a speed. From one viewpoint it is the effects of this unpredictability that impedes the consolation that are expected elsewhere.

Imagine a simple clock consisting of two mirrors A and B, in between the light heartbeat bounces. The mirrors are divided into two parts. is L, and the clock ticks every time the heartbeat of light strikes a particular mirror.

In the casing, where the clock sits very still (outline on the right) the heartbeat of light flows out in a length 2L. The time for the timer is divided into 2L by the speed of light

$\Delta t=$.

From the reference casing of a person moving at the speed v with regard to the remainder edges of the clock (chart below lower right) the heartbeat of light extends in a longer measured way. The second idea of unique relativity states that the rate of light in space is constant for each individual inertial observer and suggests a slowing of the timing of the clock from the perspective of the eyewitness moving in viewpoint. In other words, when a casing is moving with relation to the clock, the clock creates an appearance of moving more slowly. The use of the Pythagorean theory leads to the well-known predictions of relativity:

that communicates the fact that for a moving observer the duration that the clock is checking will be greater than the time in the case of the clock itself.

Due to the relative speed of communication between observers

Minkowski outline

The time UV of a check in S is smaller in comparison to Ux in S", and the time UW of the check in S is less distinct to Ux for S.

Find C's relative motion between two timekeepers synchronized An as well as B. C is in contact with An at d and B at f.

Twin mystery. One twin has to alter outline, triggering distinct moments in the twin's real lines.

A sound judgment will determine that, if time entry been impeded by a moving object and it is unable to move, then the moving object would be able to watch the outside world be "accelerated". Unlogically, the concept of unique relativity predicts the reverse.

A similar phenomenon occurs in everyday life. If Sam meets Abigail in a distance, she appears to him to be nothing and at the same time Sam appears insignificant to Abigail. Being very familiar with the implications of point of view, we can see nothing that suggests of an anomaly in this scenario. [15]

The mind is used to the concept of relativity in relation to separation. The separation between Los Angeles to New York is in fact identical to the separation between New York to Los Angeles. However when considering velocities as, one imagines an item being "really" in motion, and fails to recognize that the movement of an object is always in relation to something else like the ground, the stars or oneself. In the event that an article is moving for another, then the next is moving according to the earlier and at the same relative speed.

In the singular hypothesis that relativity is the cause of everything, a clock is observed to be ticking slowly relative to the clock of the eyewitness. If Sam as well as Abigail are riding on different trains that are in close lightspeed movements, Sam measures (by all methods of estimation) the tickers on Abigail's train to run slowly and in turn, Abigail measures timekeepers on Sam's train so that it runs slowly.

It is important to note that in each effort to establish "synchronization" in an existing framework of reference, the question of whether something occurring in one place is actually occurring simultaneously with another event taking place elsewhere is vital. Estimates are finally taking into consideration the events that are in fact concurrent. Additionally, the process of establishing the synchronization of events that are separated in space is fundamentally dependent on the transfer of data between regions that are not dependent on anyone else implies that the speed of light will be a major factor in decision of whether or not there is a concurrency.

It is a typical and genuine blue thing to consider whether, in the context of curiosity, relativity may behave naturally steady when clock C has been time-widened with respect to clock B, and clock B is also time-extended as clock C. It is through test the suppositions that are incorporated into the underlying concept of synchronization that consistency can be restored. Concurrence refers to a relation between a person in a particular frame of reference as well as the arrangement of events. Through similarity left and right, they are recognized to change according to the location of the observer because they represent the relationship. In the same way, Plato clarified that here and there are a connection to the earth, and the other is not likely to fall off the antipodes.

In relativity, fleeting-direction frameworks are created using an approach to synchronize tickers. The method is commonly referred to as the Poincare-Einstein synchronization method. A person who is onlooker and has an electronic clock emits a red flag to the sky at t1 in accordance with his clock. In a distant event the light flag is reflection back, and then touches the base again at the eye of the observer at the exact time of t2, as determined in his timer. Because the light travels each part in the same manner at the same speed moving both back and out to the observer in this scenario and the time at the moment of the light being reflect for the viewer tE is $tE = (t1 + t2)/2$. In this way, the single clock of a spectator could be used to define the direction of light that is fleeting, which is great anywhere throughout the world.

It is true that the clocks that are present throughout the other inertial casings and clocks do not appear to be at all synchronized in these areas and that's the principle of the relativity of synchronization. Because the sets of supposedly sync-synchronized minutes are distinguished by different observers, each could view the other clock like the moderate one with no relativity naturally conflicting. Symmetric time expansion occurs as directions frameworks are set up using this method. It's an effect that causes a clock to be running more slowly than the clock of one's own. Eyewitnesses aren't aware of their own clock to be affected, however, they might find that their clock is perceived to be affected in a different direction.

This symmetry is evident in this Minkowski chart (second image of how to use the privilege). Check C resting on an inertial edge S' is in contact with time A at d, and clock B is at the f (both sitting in S). The three timekeepers at the same time begin to count within S. Its worldline A is the ct pivot, the worldline of B that meets with f is parallel to the ct hub, and that of C's worldline is ct''- hub. Every time d occurs in S are recorded on the x-hub as well as in S it is on the x hub.

The most efficient time possible between two events is shown by a clock during both occasions. The clock is always invariant, i.e., in every inertial edge, it is agreed that this time will be displayed by the clock. Interim df is therefore the most efficient time possible for clock C and is less in relation to the directions times ef=dg of the checks B as well as An S. On however, the proper the time of ef in B is lower in relation to time if it's in S', because the event e was recorded in S at the time of writing at the time of i due to the relativity of synchronization. This is much ahead of when C started to begin to.

It can be realized that the greatest possible time between two events measured by an unaccelerated timer that is present at both times as opposed to the synchronized direction time that is measured on all other inertial edges is always the smallest that exists between the two events. However the interval between two times can be compared to the most efficient timing of the tickers that have been accelerated on each time. In all possible times between two instances the optimal time for the clock that is not accelerated is the maximum and is the solution to that twin dilemma. [16]

In this manner, the duration of the clock's clock cycle for the moving clock is seen to expand and it is deemed to be running "moderate". The magnitude of these differences in daily life, where there is a v c regardless of the possibility of space travel, aren't sufficient to produce noticeable time-widening effects and the effects of such tiny amounts are able to be safely ignored in general terms. It's only when an object is moving at a rate of 3000 km/s (1/10 the speed that light travels) that the expansion of time becomes to be crucial.

Time-span widening due to Lorentz component. Lorentz component was predicted in Joseph Larmor (1897), regardless of the fact that electrons are who are in a circle around a central. In this manner, "... singular electrons depict the different parts of their circle with times that are shorter than the general framework in the " (Larmor 1897). The time-enlargement of the size in comparison the (Lorentz) component has been deemed to be a possibility in the image below.

The length of time is increased due to appealing energy and movement

Day by day time extension (addition or misfortune in the event of the value is negative) in microseconds as a portion of (round) circle of r = rs/re where rs is the satellite circle sweep and re represents the thermal Earth's span calculated with the Schwarzschild measurement. If r is 1.497 (Note 1), there isn't any time expansion. Here , the effects of gravity and movement extend out. ISS space explorers are flying beneath while GPS as well as Geostationary satellites hover over. [1]

High precision time keeping and low earth circle satellite following and pulsar timing are all applications which require the consideration of the connected effects of movement and mass in the process of extending time. The most useful illustrations are based on those of the International Atomic Time standard and its relationship to the Barycentric Coordinate Time standard utilized to determine interplanetary frame references.

Time widening and relativistic impacts for the planet's closest system and the Earth can be accurately displayed by the Schwarzschild answer to The Einstein Field Conditions. In the Schwarzschild measure, the dtE interval is calculated by [18][19]

Confirmation of the exploratory process:

It has been tested at various times. The standard work being conducted in the field of molecule quickening agents from the 1950s, including CERN. CERN is a constant continuous test of the time enlargement of the relativity. The tests that are specific to CERN include:

Speed time test to determine the width of the test

Ives as well as Stilwell (1938 and 1941). The stated reason for these tests was to test the time expansion impact predicted by the Larmor-Lorentz ether theory due to the movement of the ether using Einstein's suggestion that Doppler reflection on channel beams can provide an accurate study. The tests examined the Doppler motion of radiation that is discharged by cathode beamswhen observed directly in front of them and in particular behind. The frequencies of the high and low frequencies observed did not match the conventional values determined from Einstein (1905) using the Lorentz shift, which occurs in the case of a source that is run moderately due to it's Lorentz element.

Rossi and Hall (1941) studied the amount of inhabitants living in large beams of muons that were created from the top of a mountain up to the was adrift level. Even though the travel time of muons from the top of the mountain to its base is just a couple of muon half-lives and the muon test at the base was slightly less. This can be explained by the fact that the increase in muons was attributed to their speed with relation to the experimenters. In other words, the muons were growing around 10 times faster than if they had been very still for the participants.

as derived by Einstein (1905). [22] For? = 90deg (cos? = zero) the value reduces to fdetected =?. This decrease in recurrence of the source moving can be explained by the time-widening effect and is often referred to as"transverse" Doppler impact. It was also predicted by relativity.

Time widening in 2010 was observed at a rate of less than 10 meters each second, using optical nuclear timekeepers that are connected with an optical fiber that spans 75 meters. [23]

Time-widening tests for gravity

In 1959, Robert Pound and Glen A. Rebka observed the extremely tiny gravitational red motion in the recurrence in light transmission in lower elevations in the Earth's gravitational field, which is moderately more significant. The results were within 10% within the predicted values from general relativity. The year 1964 was the first time Pound as well as J. L. Snider determined an outcome within one percent of the quality predicted by gravitational time dilation. [24] (See Pound-Rebka test)

In 2010, gravitational time-widening was detected on the globe's surface, with a height distinction of just one meter using optic nuclear clocks. [23]

Time and speed of gravitational acceleration testing of impact consolidated

Hafele and Keating In 1971, Hafele and Keating used nuclear caesium timekeepers to fly both west and east around the globe, in business aircraft in order to consider the time that passed against the clock that was based on the U.S. Maritime Observatory. Two inverse impacts could have been the most significant element. The clocks were believed to age faster (demonstrate the greater amount of time passed) in comparison to the standard clock because they had a greater (weaker) gravitational field throughout the event (c.f. Pound-Rebka test). However, also as a contrast the tickers that moved were believed to age much more rapidly because of the speed of their journey. Based on the actual flight paths of each trip it was predicted that the tickers flying with reference and contrast times from the U.S. Maritime Observatory, would have lost 40+23 nanoseconds during the eastbound trip and would have grown by to 275+-21 nanoseconds in the westbound outing. Regarding the nuclear size of the U.S. Maritime Observatory, the flying tickers dropped 59+-10 nanoseconds in the eastbound trek and increased 273+7 nanoseconds in the westbound outing (where the mistake bars refer to the standard deviation). In 2005 The National Physical Laboratory in the United Kingdom reported their constrained version of this test. [2626. The NPL test was different from the original in that the caesium timekeepers had been taken on a less arduous journey (London-Washington, D.C. return) but their timekeepers proved to be more precise. The test results are within 4percent of the expected values of relativity, within the uncertainty of the estimates.

The Global Positioning System can be seen as a continuously testing process in both the special or general relativity. The timekeepers in the circle are adjusted for both general and extraordinary relativistic effects on time, as shown above, meaning that (as seen from the earth's surface) they are running at the same pace as tickers in the atmosphere of Earth. [27]

Muon lifetime

A relationship between muon lifetimes at different speeds is possible. In the research facilities, muons with moderate energies are generated as well as in the surrounding environment speedy moving muons are displayed with incredible beams. The muon's lifetime is as in line with the estimate of the research facility at 2.197 Us, we can see that the life span of an unimaginable beam that is delivered muons moving around 98% the speed of light is five times more, in line with the observations. The muon stockpiling ring of CERN the life span of muons that were circling around with? = 29.327 was discovered to be extended up to 64.378 us, which confirms the time-span expanding to an accuracy of 0.9 + 0.4 sections per 1,000. [29] In this study,"clock "clock" refers to the duration that is taken by processes which cause muon rot. these processes take place in the muon's motion in its "clock rate" that is slower than the laboratory clock.

Space flight

A time extension will allow passengers in a swiftly moving vehicle to go farther into the past, while developing almost nothing, given that their speed slacks off the on-load section and opens and off-load time relative to the time of a viewer. The clock on the boat (and in the context of relativity, any person who goes with it) is less affected in time than the clocks of the people watching around the globe. At sufficiently high speeds, the effect is significant. For instance, the time spent traveling could be equivalent to ten years back at home. In all likelihood that a constant one g increase in speed could allow us to travel across the entire known Universe in a single life. The space explorations of the spacecraft could return to Earth thousands of years later. An example of this notion was portrayed in the book Planet of the Apes by Pierre Boulle.

One more likely use for this potential impact is to allow people to travel to nearby stars without spending their entire lives on the same boat. However that, any use of the speed of time expanding during interstellar travel requires the use of a innovative, efficient method of energy. The Orion Project is the primary effort to advance this notion.

The current space-flight technology has important hypothetical boundaries considering the issue of down-to-earth that a growing amount of energy is required to provide the acceleration of space flight as a specialty methodology the speed of light. The risk of collision with a small amount of space flotsam and jetsam as well as other particles is yet another common sense limitation. In the couple of seconds regardless there is a time increase, however it is not enough to have a significant impact on space travel. Explore spacetime regions in which gravitational time enlargement occurs, for instance within the field that is a gravitational black-hole, however beyond the visible skyline (maybe in a hyperbolic direction that leaves the field) and you could also get results that are consistent with the current theories.

Time expansion in a constant force

In the realm of exceptional relativity, the time expansion phenomenon is the most commonly depicted scenario when relative speed is constant. In any case, Lorentz conditions allow one to establish the legitimacy of time and the development of spacetime in the case of a spacecraft associated with a power of every unit of mass relation to an object with uniform (i.e. continuous speed) motion, equal to g throughout the estimation period.

Give t the opportunity to become the time in an inertial casing, hence"the rest edge. Give x an opportunity to function as a spatial facilitator and let the course of the constant increase in speed and the speed of the spacecraft (with relation toward the remainder edge) be in line with the hub's x. In accepting the spaceship's location at the time $t = 0$ with $x = 0$, the speed is v0 assigning the associated contraction

The spacetime geometrical model of speed dilation

The red and green spots in the movement refer to spaceships. The vessels of the green armada's have zero speed relative to one another, and so when the tickers are on loading up each ship, in the same amount of time. passes by with respect to each other. Moreover, they could create a plan to ensure synchronization regular armada timing. The vessels belonging to"the "red armada" are traveling at the same speed of 0.866c similar to that of that of green.

The blue specks are the beats of light. A single heartbeat of light between two green boats requires about two minutes of "green time" 1 second for each stage.

When viewed from the viewpoint of reds time of the heartbeats that they trade between each with each other equals one second "red time" for each stage. When viewed from the viewpoint of the greens they're red. Their sequence of trading light heartbeats travels in a slanting manner which is about two light-seconds. (As as seen from the green perspective, that is 1.73 (*sqrt 3 scriptstyle sqrt3) light-seconds in separation every the green clock.)

One of the red boats emits an ethereal heartbeat to the greens every second in the red time. These heartbeats are recorded from the green boats of the armada using two-second intervals recorded by green times. What isn't mentioned in the liveliness is the fact that all areas of science are inclusive. The heartbeats of light which are transmitted by reds with a certain frequency as measured in red time are received with a lower recurrence, according to the identifications in the green armada, which are compared against green time and vice versa.

The liveliness varies between the green viewpoint as well as the red view to highlight the symmetry. Since there isn't any thing as a clear motion within relativity (as is also the case with Newtonian mechanics)

The red and the green armada are regarded by view by virtue of being unmoving within their own frame of reference.

Remember that the repercussions of these collaborations and calculations reflect the actual state of the vessels as they emerge from their environment of relative motion. It's not an easy element of the technique to estimate or correspond.

Chapter 9: Major Scientific Study Launched
Robert Bigelow intended to find the root of the issue that was going on on the Skinwalker Ranch, and he was determined to find the cause through the written word and using the solid scientific principles of mainstream science and a logical approach.

To accomplish this he created his organization, the National Institute for Discovery Science (NIDSci) which is an organization which focuses on research into "fringe phenomenon" or paranormal phenomena, specifically concerns related to issues related to the UFO issue. It is important to note that NIDSci wasn't just an amateur organization, like MUFON or any of the other paranormal investigation 'clubs' which are gaining popularity over the past year. NIDSci was run by researchers from the mainstream who had doctoral degree in university education with world-class credentials.

Bigelow had the money to buy the most beautiful.

The researchers set about to wire all zones of the ranch using sophisticated technology for electronic surveillance. They put up enough automated cameras to cover all areas with sensors of all types that range from sensors for night vision to monitoring devices that use electromagnetic energy. They set up a trailer filled with infrared and computerized sensors geomagnetic anomaly readers as well as a myriad of other equipment. They kept a 24-hour staff within the premises.

The researchers conducted a geomagnetic investigation to determine if they could find a naturally occurring gravitational anomaly. They conducted a survey and analyzed every species of plant at the ranch to look whether there was any naturally hallucinogenic chemicals, like "magic mushrooms" or similar substances that could cause people to experience bizarre visions. They speculated that compounds like these might have leaked into groundwater, but all water tests proved to be normal, and no plants that could cause intoxication were discovered. They conducted interviews with hundreds of people from the region including neighbours, as well as any person who visited the ranch.

After an exhaustive analysis of the situation KLAS-TV reporter George Knapp said that it was not long before scientists started to see the bizarre phenomena themselves, such as lights, orbsand Monsters UFOs as well as a myriad of other mysterious events. Knapp revealed this to an MUFON audience at an address on June 17, 2008:

"There was an Ph.D. physicist who had been working on classified programs for a variety of government agencies and military organizations. He was out on the property along with Dr. Colm Kelleher. They saw the black object in the trees, a sort of an undefined cloud. It then begins to circle the dogs in a terrifying manner. The dogs together with them are scared. The guy the physicist begins speaking in a different voice. This voice has entered his brain, and it tells Colm that they're looking over him and that he's not welcomed there. The physicist isn't aware of the incident however, for a few days afterward the thing stayed on him and terrified his other places on the property."

Knapp was then quoted as having the account, which sounded as if it could be described as a 'portal' in a parallel universe incident.

"These are world-class scientists ... There was an additional instance where there were two people who were on top of a ridge, with instruments and telescopes, and two others below. Two guys below begin to notice this dark snowball of light yellow hanging around one foot above the ground in the middle of the grass ... the snowball is becoming larger and larger and they're talking on their walkie-talkies with the guys on the ridge. They ask "Can you spot the thing?' they respond, 'Yes, they're watching. This ball grows larger, then it expands until it appears like an underground tunnel ... and then, the guys wearing the infrared said, "Hey there's something in this thing.'

"And it is true that this huge humanoid being starts moving across the passageway from one end trying to move to the other, like it's from other. It reaches the final point of this tunnel of illumination. It rises up. It's about the height of eight foot, black, and featureless and then it leaps out. The tunnel collapses and then disappears. it runs upwards up Skinwalker Ridge ... and the guys who are there needed to change their clothing as they think it's going to chase them. However, it wasn't ... It was just poof!"

One of the most notable members from one of the NIDSci group was former U.S. Army Colonel John Alexander who graduated with an Ph.D. in sociology as well as educational sciences, and later was a commander within the U.S. Military's top Army Special Forces. The credibility of his name was quite high. That's because the person who was not an untrustworthy UFO-related nut as well as a New Age seeker. Alexander was one of the principal investigation team of Skinwalker Ranch. Alexander told Open Minds TV:

"What we found out was that the incidents (on at the Skinwalker Ranch) are real and real-time and clearly taking place. They were not figments of someone's imagination or folklore, or anything of the sort of events ... however, we're still baffled."

Alexander claimed that the "intelligence" (or what it was) at the Skinwalker Ranch seemed to revel in playing games of mind with scientists never a step behind whatever they attempted to do. At one point, Alexander said the 'intelligence was able to dismantle one of the video surveillance systems that were mounted on a pole in the edge of the pen for cattle. Alexander explained the incident in this manner on the following Open Minds TV interview:

"The cameras were situated on the top of platforms. They were fitted with PVC connectors that held them to the dirt ... There was also a lot of duct tape to hold the wires in place. Sometime later, the wires broke loose and camera stopped recording. The tape used to secure the post has now gone ... it's not even cut ... it's gone. The PVC clamps have gone. They are gone. PVC posts are pulled out. There is a piece taken out of the wire around three feet long. It is missing."

What makes the destruction on the equipment for cameras mounted on the pole more intriguing is the fact the fact that other cameras were located in the vicinity and another camera could have seen the setting-up of the demolished camera pole. The camera that was 'watching a camera' was not able to record anything happening during the time that the act was completed.

Alexander is fond of calling the intelligent force operating at the Skinwalker Ranch "a trickster" since it was always one step ahead of any effort by scientists to collect information about it. It was also evidently fond of manipulating the minds of those working on any type of task or experiment, whether it was scientific or ranching.

A Night Watchman is entangled in the Parallel Universe

Many strange stories relating with the Skinwalker Ranch were reported and gathered from a myriad of sources, they have almost no limit.

Another highly compelling story of parallel universe activities in the Skinwalker Ranch is told by the hand of the writer Erick T Rhetts's novel, Lost on the Skinwalker Ranch.

Rhetts tells the readers that his book tells of real events that were experienced by an individual who was employed as night watchman and security guard on the ranch.

The security guard was initially able to relay about his experiences to Rhetts under the condition that to be kept anonymous. One reason for the need to keep his story secret is the fact that Bigelow demanded employees to agree on a confidentiality contract regarding anything that transpired on the premises. Also, they were not permitted to conduct interviews with the press.

After quitting his post, the night watchman departed for the United States to live as an expatriate in Peru and decided to let his account of the Skinwalker Ranch made public.

Like all those who spent a lot of time in the Skinwalker's land and the Skinwalker, the former security guard was a victim of plenty of terrifying and bizarre encounters. They also saw frequent reports of glowing orbs ghostly humanoid figures wandering around the dark landscape in the evening and a handful of UFO sightings, as well as numerous terrifying encounters with cryptoid beasts like giant wolves as well as standing 'wolfmen'.

It was a brief encounter with two of these humanoid wolf-like creatures that led to the most terrifying of situations and a flimsy escape into an interdimensional portal and a mind-boggling journey into a different world or maybe an alternate timeline scenario.

It all started one night while the security guard was on patrol. Security guards were on an errand in darkness while walking with care through shrubs and the rocky desert-like terrain of the ranch. He observed what initially appeared like two men walking toward him.

As he aimed a strong illumination on the two figures and was shocked, it was not two people but two wolves standing upright on two legs, much like men.

The two wolf-like men were slim and tall and looked almost emaciated. As they came across an officer in the dark, they started sprinting towards him, and it was evident that their intention was not to engage in a conversation with a friend or to seek directions. Guard was persuaded the wolves were humanoid and had plans to kill him, and they turned and fled.

The guard ran away towards an old, abandoned house which had been within the grounds for years. It was essentially a rotten out house's frame that had a roof leak that was barely up and running, with nothing more than dirt as flooring. The doors and windows have been demolished. The building would have made perfect props for the setting for a haunted home in the set of a Hollywood film, as it was known for its bizarre incidents that occurred within it. Others had also claimed that strange looking creatures such as ghosts, ghosts and other creatures were seen entering into the cabin but were not observed to return.

While the security guard wasn't interested in entering the haunted house, he knew that he was forced to, because the werewolves of the dark are now at his heels.

After a race through the darkness night, the guard made it to the cabin, and plunged into the empty doorframe but as he entered the cabin it felt as if it was falling, however instead of hitting the ground, he felt himself sliding - and then sliding through what seemed like an empty, dark space.

After an undetermined amount of confusion and time, the man was seated on the ground. He it was clear that he was not in the interior of a cabin that had been abandoned in Utah. The guard found himself inside what was believed to be an enormous underground cave.

He was stunned and started to search around, trying to discover a way out. It was interesting to note that he wasn't totally dark. A dim yellow light that was emanating from the distance and all over produced an ambient glow which allowed him to see only enough to see a tiny passageway. This path was clearly designed by a human being, as it was comprised of steps that were moving upwards.

While he walked through the narrow stairwell the cave was filled with paintings of strange characters and bizarre images on cave walls. After what appeared to be an unending time in the cramped confines that were the cave's walls, the man finally left and was astonished. The relief was only temporary however, as the cave was now the bottom of a bizarre valley that was surrounded by the massive peaks of stony rock.

In the high canyon's walls, the long-lost security officer was astonished by the cave dwellings and people milling around within them the caves he could clearly observe human figures. The caves were lit by warm yellow glow from campfires. Against these were visible the silhouettes of people dressed in primitive ways.

There were other adventure stories in the future - suggest Erick T Rhett's book to read the entire story - however, the man did find his way back to share his tale. It is evident that the story is full of the sounds of an inter-dimensional slip the portal, which landed an uneasy night watchman in another world or maybe a different place.

Sky Portals

Additionally, frequently reported in the vicinity within Skinwalker Ranch Skinwalker Ranch and in areas near by are what appear as energetic open-air spaces that appear to be openings in the skies. The book, Future Esoteric: The Unseen Realms, author Brad Olsen writes:

The most bizarre of the phenomenon (on one of the ranches, Skinwalker Ranch) are the occasional sighting of a large, orange portal in the sky, that seems to open up to another dimension. In the night, blue skies is visible through the portal as well as black cars have also been seen entering and exiting this portal ... Alongside to magnetic anomalies and teleportation it also seems to be an intrusion into alternate reality. Parallel universes or gateways to higher dimensions, are linked with the portal in orange. The portal, in addition to only a few others that have been known to be in existence, could be the path for cryptozoology creatures to enter and exit into our 3D reality.

A Mystery that is a Puzzle Unsolved

Whatever is going on at this Skinwalker Ranch, any attempt to pinpoint it and quantify or identify it or even provide a straightforward idea about it has completely failed. A lot of the most brilliant minds have been there and there is a constant stream of curious visitors. According to the Colonel John Alexander, one of the principal investigators of the Bigelow's NiDSci team, told me:

"Anybody who thinks they comprehend the real nature of the phenomenon simply fooling themselves."

The Skinwalker phenomenon was a snare to the greatest efforts, high-tech and other, of the amazing scientists from NIDSci. Today, NIDSci is no more. Robert Bigelow pulled the funding and shut down his beloved paranormal investigators. The group disbanded in 2004.

The only way to know something is those Native American elders who still reside on the land. They believe that the bizarre phenomena and sightings have been a normal aspect in the land for hundreds of years. It is true that the Ute as well as the Navaho may not fully comprehend the creatures and skylights that inhabit Skinwalker. However, they know that they are there and know how to be with it for centuries to ending. They also know:

It is best practice to not talk about what transpires at the Skinwalker Ranch.

Within his novel, Skinwalker and Beyond, the author Ryan Burns writes:

Similar to other parts of Native American mythology Tribal members guard the secrets of the issue. But there is a important distinction, that is the issue of Skinwalker is dealt with the greatest respect. It is not something that should be discussed. Many people believe that even the mention of the Skinwalker word can trigger the very thing that is being talked about ... To truly comprehend the background of Skinwalkers, it is essential to know the history of the culture that spawned the Skinwalker.

However, with the development of modern technology and the growing interest in UFOs as well as space travel, and maybe even interstellar space travel, inter-dimensional travel and time travel, the likelihood that there are doors that exist in this space will always be a form of catnip for individuals who are interested or simply want to be aware. They're eager to discover and are always keen to discover those hidden ways that can open up to new, perhaps fantastical worlds.

Therefore, the fascination with the Skinwalker Ranch is unlikely to go away, especially since the internet and media has spread the fascinating tales of this place across the globe. If there are potentials and unanswered mysteries there are always those who are willing to risk everything and try their best to find out what's possible find at The Skinwalker Ranch.

A Third Night

Following the dramatic events that took place the night before on the second night of the encounter the UFOs returned a third time in Rendlesham Forest - and this is perhaps the most thrilling however also the most controversial event of three days. festival.

While there are numerous witnesses, including local police, military and everyday people, the most shocking tale comes from an individual who was an untrained private at the time. Airman First Class Larry Warren, who was only 18 years old.

Warren was assigned to work for duty at RAF Bentwaters three weeks earlier. He had requested to work at the RAF Bentwaters since his father worked at the U.S./British base before him. Warren was educated to work as a member of the security team at Bentwaters and that night was working as a guard at the gates , when bizarre events began to occur.

Airman Warren didn't know the situation at first, but he was removed from guard duty and instructed by his superiors to be part of the team heading to Rendlesham Forest. What Warren and a lot of others in the military then began to see was truly amazing.

The soldiers could clearly see moving lights throughout the forest. As they made their way through the woods they saw numerous objects. One was a triangular structure that had a red light on its top with a line lights running down the bottom. it appeared to be similar to the object Burroughs and Penniston saw the first night, and that the Colonel Halt and his crew were able to see on the second night. They also observed the sphere glowing in red that "exploded" in the direction of them, sending the sphere of light with red or particles into the ground below. Warren stated that no one in his group was injured in any way from this explosion. There were a myriad of other stars in the sky too.

When the glowing red light that exploded, Warren said the unit changed and became an object like the one seen earlier. This time, it was the triangular UFO, with an red light on the top and rows of light across the bottom. While the stunned men stared in awe, and unable to comprehend what they saw and what was happening, the most bizarre events of the night were set to take place.

Four years after, Larry Warren told it this to TV journalists from CNN:

"At this point, we saw people or whatever you wish to label them come from the ship ... There no doors on the ship. These things went to the side ... they were all three. They had lights around them. They emerged, and as they remained in the middle, making things more complicated they actually floated. They were beings. They did not make any threatening gestures. They didn't communicate with me in any way. The commander of our base was present ... Lieutenant. Colonel. Gordon Williams."

Gordon Williams initially denied he was present at the scene at the time of the event and later declared Warren as a deceiver. But, the documents that were discovered years later proved that Gordon was indeed there on the night of the incident, which suggests that Warren was not lying at the end of the day.

Warren said that a senior officer had arrived and tried to establish communicating with the alien beings. There was no exchange that took place. In actual fact, Warren described the human-to-alien confrontation as a standoff. He then said he returned to the base, while events continued to unfold behind him. He arrived back about 4.00am. He claimed that the UFO was within the area at the time he went to leave.

Following the incident, Warren said he was "debriefed" and was forced to sign a document stating that he did not see anything other than the lights of the sky. He was also given the strict orders not to talk about the incident to anyone else including the other airmen in his unit, not even those who were witnesses to the whole incident. In essence the order was to erase everything the event had shown him and was told that ignoring the event could be a grave risk including being dishonouredly discharged from the military to perhaps even the possibility of a prison sentence.

But, Larry Warren began speaking about his experience shortly after he finished his service and was awarded an honorable discharge, even though there was an obligation under law to remain silence. He began speaking to journalists and film crews, as well as recounting his story during UFO conferences. Some military personnel have claimed Warren made up the whole story up, while others have supported his account. The military did not take any legal actions against Warren however, they did officially label him a deceiver.

The fact is there is no doubt that the U.S. military made strenuous efforts to ignore the facts to minimize the story and conceal it. That the military was active in hiding the truth was the conclusion reached by CNN journalists who broadcast an investigative documentary about the Bentwaters incident in 1984. Many documents released later under Freedom of Information Act Freedom of Information Act all however confirm that an official cover-up of the incident was in place and was kept.

An Unfolding Story Time Shifts and Time Travel

What we have provided above is merely a brief overview of the Bentwaters-Rendlesham Forest UFO event which, as we have said, is a story of enormous complexity, and one that has only grown more rich and variegated over the years as additional witnesses have come forward, adding details and further eyewitness accounts to the narrative. (For instance mainstream journalists as well as UFO researchers have later uncovered civilian witnesses who also were in close proximity to those in the Rendlesham Forest that night, and they all reported seeing strange lights in the sky. It was also reported by a local veterinarian and motorists who were passing along a nearby highway.)

Conclusion

It's been more than sixty years since the physicist Hugh Everett III first proposed his Many Worlds Interpretation and scientists have been discussing it since. The reason for this is that the concept of "Many Worlds" is amazing. It's not just that our universe or reality dimension not the only one of the dimensions of reality, it's only one of the many of parallel universes. And in fact the 'other worlds' are located with ours. Perhaps the thinnest veil-like cloaks separates us from the other worlds.

Many scientists believe that they believe that the Many Worlds theory carries too many metaphysical implications. However the results of a study conducted on the scientists from around the world ten years ago found that 68% are now in agreement on they believe that the Many Worlds Interpretation is either "true" or "mostly accurate'.

www.ingramcontent.com/pod-product-compliance
Lightning Source LLC
Chambersburg PA
CBHW050404120526
44590CB00015B/1818